STRUCTURE AND PROPERTIES OF INTERMETALLICS IN PRE-TRANSITIONAL LOW-STABILITY STATES

STRUCTURE AND PROPERTIES OF INTERMETALLICS IN PRE-TRANSITIONAL LOW-STABILITY STATES

A.I. Potekaev

A.M. Glezer

V.V. Kulagina

M.D. Starostenkov

A.A. Klopotov

CRC Press is an imprint of the
Taylor & Francis Group, an **informa** business

Translated from Russian by V.E. Riecansky

CRC Press
Taylor & Francis Group
6000 Broken Sound Parkway NW, Suite 300
Boca Raton, FL 33487-2742

© 2021 by CISP
CRC Press is an imprint of Taylor & Francis Group, an Informa business

No claim to original U.S. Government works

Printed on acid-free paper

International Standard Book Number-13: 978-0-367-48972-4 (Hardback)

Contents

Foreword

At present, scientific and technological development poses the challenges of materials science to create fundamentally new materials with unique properties. It is known that the physical and mechanical characteristics of condensed matter are to a large extent determined by their structure. The use of experimental and semi-empirical approaches has objectively exhausted itself, and the semi-empirical representations of classical materials science are progressively losing their predictive capabilities with an increase in the number of components in the system and stricter requirements for the operating and functioning conditions. The conceptual ideas that existed up to now on the nature of materials and methods for constructing their properties and predicting behaviour under extreme conditions have practically exhausted their capabilities. In this regard, there is an urgent need to create effective methods for predicting the structure and properties of condensed systems based on new physical concepts and approaches.

Of particular importance are the two latest problems in the physics of the condensed state of systems. First of all, it is the physics of the formation, behaviour and properties of systems with a wide range of states of metals, alloys, semiconductors, all of whose states are in the vicinity of the stability boundary. A common feature of the systems under consideration is their low stability with respect to influences. In this case, a low-stability state of a system is understood, as a rule, to be such a state near a structural–phase transformation in which anomalies of the structure or properties are observed. Another important problem is the unusual structure and unusual behaviour of nanosystems, i.e. systems with nanoscale elements. For the latter problem, the following is well known. If the range 1.0–0.1 μm is a complex technological barrier, since it requires a change in technological equipment, then the range 0.1–0.05 μm is a fundamental physical barrier, beyond which all physical properties of a solid change sharply. Visual images and familiar views and models lose their strength.

Both of the above problems are today among the most relevant and significant in the physics of condensed systems. Naturally, in these cases, the structural phase states do not correspond to traditional concepts. First of all, structural defects under these specific conditions become integral elements of the structure, interact with each other, and this interaction has a significant effect on the structure and properties of the condensed system itself. It should be specially noted that in both cases the density of structural defects (defects in the traditional sense) is very high, therefore they cannot be considered as isolated, and it is necessary to study the system of interacting defects under conditions of alow-stability state of the material. This in itself is far from a trivial task, especially considering the fact that an important role is beginning to be played not only the concentration of defects but also their symmetry, the nature of the interaction, the plane of occurrence, the type and magnitude of the external influence, and much more. Against the background of the state of a condensed system weakly resistant to external influences, the role of the interaction of structural defects acquires a particularly important, and often decisive, significance for the structure.

This book is devoted to the fundamental, primarily physical, aspects of the structure and properties of the state of a condensed system of perspective materials that with low-stability withrespect to to external conditions.

So, on the way, dear reader. I hope it will be interesting.

Introduction

Metallic materials with a BCC structure are currently widely used because of their unique physical and mechanical properties. Two groups can be distinguished among them, whose representatives have excellent structural-phase features and behavioural patterns, especially in the high-temperature region of low-stability pre-transitionalal states: traditional alloys (e.g., Cu–Zn system alloys) and intermetallic compounds (e.g., Ti–Ni and Ni–Al). A large number of works have been devoted to pre-transitionalal low-stability states preceding structural–phase transformations in alloys, for example, in those based on titanium nickelide. Manifestations of pre-transitionalal features, reflecting well-defined low-stability states, are characteristic of a wide range of both metallic [1–18] and non-metallic systems [19–26]. The notions of pre-transitionalal low-stability states rely on a large number of studies, the most notable among them discuss the functional intermetallic alloys of the Ti–Ni system.

TiNi-based alloys are most effective for designing materials with shape memory effects (SME), which have found a wide application [3]. The choice of alloys for developing functional materials with SME mainly depends on the set of optimal properties. Among other things these are the temperature ranges of martensitic transformations (MTs), the value of the SME, the value of the martensitic shear stress, the deformation and temperature cyclic resistance, etc. [3, 25–27]. In order to form the required optimal target physical and mechanical properties, it is necessary to study traditional alloys (e.g., alloys of the Cu –Zn system) and intermetallic compounds (e.g., alloys of the Ti–Ni and Ni–Al systems).

It is now understood that pre-transitionalal phenomena result from the manifestation of low-stability states of both crystalline and amorphous solids. A low-stability pre-transitional state of a system is understood as such a state of the system near the structural–phase transformation in which anomalies in the structure or properties are observed [see, e.g., 28–50]. The objects under consideration in this work are the alloys and intermetallic compounds that undergo structural-phase second-order transition or a transition similar to it. From a physical point of view, based on a new understanding of the state and properties of the system, the traditional phase transition point represents a range of values of the parameter controlling the transition. In this interval, the state of the material is structurally unstable with respect to the influence of small changes in the controlling parameter. In addition, it should be noted that pre-transitionalal phenomena are accompanied by a mixture of phases with similar structural–phase states, which vary with insignificant changes in the external conditions [1–18].

In these pre-transitional states there are many phenomena that are not clear at the present time. For example, pre-transitional phenomena do not always precede the transition, which then actually occurs with the formation of a final crystalline structure. The pre-transitional state of the high-temperature symmetric phase, contains 'signs' of many transitions. In the transformation, the initial crystalline structure is converted to the final structure, not directly, but through a whole range of intermediate stages. An example of the most common cases of this kind is the formation of long-period and polytype structures, both one- and two-dimensional structures In this case, the transitions can occur via different mechanisms, both diffusion-free (FCC↔HCP, martensitic, etc.) and diffusion-induced mechanisms such as superstructures such as Cu–AuII, Cu3Pd (M), Laves phases, etc. are formed [1–18, 51–59].

All this allows us to assert that condensed systems undergoing phase transitions are interesting in that their crystal lattice in the transition region has the features characteristic of pre-transitional processes. It is known that structural phase transformations are one of the most difficult to describe phenomena. This is due to many reasons, we will mention some of them below.

First, the transition from one crystallographic modification to another occurs under conditions of a limited number of degrees of freedom of the system.

Secondly, the integration of the old and new phases during structural–phase transformations is commonly difficult, which leads to significant stresses at the interface. Therefore, the fields of interfacial stresses affect the material at large distances from the interface (of the order of several nanometers) and can contribute to significant changes in the physical properties of the phases both near the interphase boundaries and far from them. It should be borne in mind that the stresses developing in the local regions near the interphase boundaries contribute to the generation of crystalline structure defects and their movement.

Third, in addition to the processes related to volumetric diffusion, diffusion processes due to boundary diffusion influence the polymorphic transformations in the alloys.

Fourth, the entire transition can be accompanied by strongly pronounced phenomena of weak stability or instability of the crystal lattice to certain influences (shear stresses, thermal fluctuations, etc.).

The degree of deviation of the system from the equilibrium state can be very different. In this case, two main polymorphic transformations (kinetic–morphological type) can occur: normal (the so-called thermofluctuation) and distortion (shear or martensitic) [12, 60]. The implementation of a specific transformation mechanism depends on the stress relaxation in the system, which can occur in various ways. Transport phenomena or shear processes in the zone of transformation.

It should be borne in mind that the systems experiencing phase transitions are, strictly speaking, nonequilibrium. This allows, based on the phenomenological approach, to differentiate between the modes system's transition to new states depending on the ratio of the relaxation times of the internal and control parameters [60, 61]. Internal controlling parameter in ordering alloys during the order↔disorder phase transitions may be a long-range order parameter [62]. For shear phase transitions, the internal controlling parameter can be the value of cooperative atomic displacements from the initial state at the sites in the crystal lattice nodes [15, 63]. The above-mentioned emphasizes that the problem of low stability or instability of various structures is very complex, therefore, it is necessary to study both the structures and states themselves, especially in the vicinity of structural–phase transformations, and their mutual transformation under the influence of various factors. Therefore, the fundamental studies revealing the physical essence of structure formation processes under thermocyclic influences in the alloys exhibiting a pronounced low stability or instability of the crystal lattice are currently very relevant. The regions of structural–phase transformations are of particular interest, taking into account their specific features of structural–phase states, the cycling temperature range, the nucleation mechanisms realized in this case, the interaction of crystal structure defects and their influence on the structural–phase state and transformation paths.

The purpose of this book is to identify general patterns of thermocyclic impacts on the changes in structural phase states in functional alloys with low stability or instability of the crystal lattice in the region of structural phase transformations. Since without understanding the physical laws of the influence of thermomechanical impacts on the properties of alloys it is impossible to develop scientifically proven technological processes for material treatment, the book widely uses the results of multi-year research by the authors of the book.

The parts and products made of TiNi-based alloys are often exposed to external stress and temperature during operation. The greatest challenge for material research is the stability of the properties of the material used. It is known that the critical temperatures and critical stresses for the B2↔B19' martensitic transformation in NiTi are very sensitive to cycling both in thermal and mechanical cycling [3]. The process of thermal cycling through the region of martensitic transformations in alloys is accompanied by the build up of defects [3, 15, 16]. The study of structural–phase states corresponding to the changes in the physico-mechanical properties of intermetallic compounds in the region of transformations is necessary for gaining an insight into the nature of the influence of thermal cycling on the properties of functional alloys.

It is currently accepted that alloys with shape memory effects (SME), depending on the conditions of operation, are conventionally divided into two groups. The first group is made up of the alloys with SME, in which a single operation occurs in devices (thermomechanical couplings, medical stents, prostheses and clamps, etc.). The second group consists of the alloys means for the involving multiple manifestations of SME (thermal drives, sensors, working fluids of heat engines, etc.). One of the main parameters in these alloys is the stability of their properties under thermal cycling in the MT temperature range. This parameter corresponds to the thermomechanical stability of SME alloys, which, by definition, is the ability to maintain the properties within an unlimited number of thermal cycles in the MT temperature range of martensitic transformations.

On the one hand, TiNi-based alloys are the most common SME alloys due to the combination of their unique functional properties [3]. On the other hand, in the equiatomic composition region they do not fully exhibit their functional properties. According to [3, 54–67], in TiNi-based alloys, thermal cycling in the MT temperature region of causes a change in the SME parameters. In addition, there is even a change in the staging

of the MT [68–70] and the accumulation of irreversible deformation [66, 71–73]. Such changes worsen the performance properties of products made of the SME alloys. Therefore, studies aimed at studying the effect of thermal cycling on the structural phase states of alloys and intermetallic compounds are relevant.

During thermocycling of TiNi-based alloys, a change in the SME parameters is associated with an increase in the density of defects and their motion and, as a result, their influence on the intensity of nucleation of martensitic crystals and on the mobility of interphase boundaries during MTs [74–77]. Different approaches and methods are used to increase the stability of the properties of SME alloys [3, 64, 75, 77–79]. They often complicate the process of accumulation of irreversible deformation during thermal cycling of TiNi-based alloys.

The concepts on the mechanisms of accumulation of irreversible deformation during thermal cycling reported in the literature are sometimes debatable. A number of works showed that irreversible deformation accumulates during cooling in the course of a direct MT [64, 71, 75, 80]. Other authors argue that irreversible deformation accumulates during heating during the reverse MT [81–83]. On the one hand, the current variety of different ideas about the mechanisms of accumulation of irreversible deformation during thermal cycling is associated with the complex nature of the processes that occur during MT. On the other hand, this is due to the sensitivity of MTs to the heat treatment and the chemical composition of the alloys. In this case, the specific features of the thermoelastic MTs in the alloys of this class are not taken into account, namely, the presence of low stability and instability of the crystal lattice in the MT region. Therefore, from a practical point of view, it is critical to obtain the data on the general laws of changes in the structural-phase states in the alloys undergoing transformation. Despite the fact that a lot of attention has been paid to this issue in the literature, the mechanism of accumulation of irreversible deformation during thermal cycling in TiNi-based alloys is not fully understood. Therefore, it is necessary to study in detail the structural–phase features of low-stability pre-transitional states and properties in the vicinity of transformations, which can become fundamental when creating an adequate model describing structural-phase changes during thermal cycling in alloys with low stability and instability of the crystal lattice in the region of structural-phase transformations.

Chapter 1 of the book contains the results of studies of the features of pre-transitional low-stability structural-phase states of BCC alloys during thermal cycling by the Monte Carlo method of the traditional CuZn alloy and NiAl intermetallic compound) in various cyclinc scenarios: during one or several thermal cycles, in the presence of complexes of planar defects – (antiphase boundaries) and their interactions.

It is shown that in all cases, as a result of each heating and cooling cycle, a peculiar hysteresis is observed whose presence indicates the irreversibility of the processes taking place in the material, which implies a difference in the structural phase states during heating and cooling. It was found that structural phase transformations in the heating and cooling stages occur in different temperature ranges.

In these intervals, the thermodynamic stimuli for the realization of one or another structural phase state are very small, which can be traced both on the dependences of the configurational energy, long-range and short-range order parameters, and from the changes in the atomic structure and distributions of structural-phase states. Both ordered and disordered phases, a certain set of superstructural domains, are realized simultaneously. This means that in the vicinity of the disorder–order phase transition there are low-stability states are observed.

It is shown that the influence of planar defects (antiphase boundaries) on the disordering process is significant up to the temperature of the structural-phase transformation. The most critical factor for the long-range order is the appearance of the defect itself, the difference in the type of antiphase boundaries (APBs) and the plane of their occurrence does not appreciably affect the behaviour of the long-range order with temperature. The type of APBs also significantly affects the structural and energy characteristics of the system at the temperatures below the phase transformation temperature. Naturally, a system with structural defects is less ordered than a defect-free system. The presence of a defect contributes to the onset of disordering of the system at lower temperatures: a decrease in the order in the alloy begins in the case of thermal APBs at a lower temperature compared with the case of shear APBs. Their presence also influences the stability of the alloy during heating. The disordering process is accompanied by blurring of the borders and their faceting.

From a comparative analysis of the features of disordering processes in the BCC system (traditional CuZn alloy and NiAl intermetallic compound) with increasing temperature in the region of low-stability pre-transitional states, it follows that if the order–disorder phase transition occurs in the traditional CuZn alloy as a result of disordering in the system, then in the NiAl intermetallic compound the long-range ordering is established as a result of structural–phase transformation. In fact, the disordering process dominates in the traditional CuZn alloy, and the structural-phase transformation dominates in the NiAl intermetallic compound.

Chapter 2 discusses the influence of various factors on low-stability pre-transitional structural-phase states of the NiAl intermetallic compound. Using the Monte Carlo method, the example of the NiAl intermetallic compound of the Ni–Al system it is shown that the processes in the BCC intermetallic compounds, which occur during thermal cycling followed by structural-phase transformations, are irreversible. The hysteresis is observed as a result of the heating–cooling cycle, which implies a difference in the structural phase states at the heating and cooling stages. It is shown that for a hypothetical order–disorder transition to be realized, the system has to be slightly overheated with respect to the traditionally understood phase transformation temperature, and likewise for a hypothetical disorder–order transition, the system has to be slightly supercooled relative to the same temperature.

During the structural-phase transformations following stepwise cooling in the alloy, two antiphase domains with the $B2$ superstructure are formed. Analysis of the atomic and phase structure of the system during cooling showed the presence of the elements corresponding to superparticle dislocations in the <100> plane and antiphase boundaries.

Analysis of the influence of various factors (concentration of vacancies, deviations of the atomic composition from stoichiometric, grain size - cell size of the model) on the structural phase states and energy characteristics during heating and cooling showed that the considered factors have a significant effect on pre-transitional low-stability structural phase states before turning.

Analysis of the influence of APB complexes (pairs of shear APBs in the <110> direction and pairs of thermal APBs in the <100> direction) demonstrated that in the region of low-stability structural-phase states, the contribution of APBs to the disordering process is significant up to the temperature of the structural phase transition.

The most critical for the long-range order in the system is the very appearance of a defect in the form of APBs; the difference in the type of APBs and the plane of their occurrence does not strongly affect the behavior of the long-range order with temperature. The presence of a defect in the form of APBs contributes to the onset of disordering at lower temperatures. In an alloy with a complex of thermal APBs, the first structural order disturbances always appear near the Al–Al interface. In an alloy with a complex of shear APBs, structural order disturbances at low temperatures are observed only in the regions of boundary intersection. The presence of antiphase boundaries affects the stability of the alloy upon heating. It is shown that the disordering process is accompanied by blurring of the boundaries and their faceting.

Chapter 3 investigated pre-transitional low-stability states in TiNi-based alloys. Using the influence of point defects and their complexes on structural transformations in the TiNi-based alloys, it has been shown that in the low-stability states of a condensed system (in TiNi, these are the so-called pre-transitional states), the interaction of structural defects can have a significant effect on structural–phase transformations and transformations objects play a very important role in phase stability in the MT region.

It has been found that the concentration dependences of the onset temperature of a direct martensitic transformation M_s in stressed and unstressed specimens are. This allowed us to conclude that the observed effect is associated with the presence of low-stability pre-transitional states in the MT region in the alloys of this class, and the structure and properties of these states depend on previous thermomechanical influences and system states.

The studies of the physical properties in multicomponent Ti (Ni, Co, Mo) alloys, the effect of annealing and thermal cycling on the MT intervals revealed that thermal cycling in the MT region in microalloyed alloys leads to an insignificant decrease in the starting temperature of MT and a noticeable increase in the area under the temperature curve of electrical resistance with saturation after the 20th cycle.

Chapter 4 discusses the relationship of structural defects and low-stability pre-transitional states, phase-structural transformations and stability of alloys. The relationship between the point, planar defects and their complexes and low-stability pre-transitional states, phase transitions, structural transformations and stability using BCC alloys as examples.

Computer simulations have shown that in the low-stability state of a condensed system, acooperative interaction of point and planar defects can lead to their ordered arrangement, and the resulting static displacement fields can both stabilize the $B2$ structure and contribute to its instability and martensitic phase transition. In the presence of a certain type of defects in the $B2$ structure, the latter is unstable to displacements of {111} planes along the direction of the type <111>. In the final structure, the displacement fields around the defects are localized, and the defects themselves organically fit into the structure of the resulting phase. Structural defects of the parent phase become natural elements of the structure of the final daughter phase. Structural defects in the low-stability state of the parent phase determine, in fact, the structure of the final daughter phase.

It is shown that in the vicinity of structural–phase transformations in CuPd alloys containing 40 at.% Pd, there are low-stability states accompanied by a number of such anomalous phenomena as the anisotropy of atomic displacements, concentration inhomogeneities, stratification, heterophase fluctuations, nonlinearities in the dependences of the lattice parameters and long-range order parameters, etc., which prepare the system for transformation.

Using tpoint and planar defects and their complexes as an example, the inheritance of structural defects by the daughter phase during structural-phase transformations in the pre-transitional low-stability state of metal BCC systems has been studied.

The authors consider it a pleasant duty to thank their colleagues A.A. Chaplygin and P.A. Chaplygin, in collaboration with whom the results reported in this book had been obtained.

Peculiarities of the pre-transitional low-stability structural-phase states of BCC alloys during thermocycling

Using the Monte Carlo method, we examined the features of thermal cycling of pre-transitional low-stability structural-phase states of BCC alloys (using a traditional CuZn alloy and a NiAl intermetallic compound as examples) in various situations: during one or several thermal cycles, in the presence of complexes of planar defects (antiphase boundaries), and interactions between the complexes of thermal antiphase boundaries.

It is shown that a peculiar hysteresis is observed in all cases as a result of each heating and cooling cycle. Its presence indicates the irreversibility of the processes taking place in these materials, which implies a difference in their structural-phase states during heating and cooling. It is found that the structural-phase transformations in the stages of heating and cooling occur in different temperature ranges. In these intervals, the thermodynamic stimuli for the realization of one or another structural-phase state are very small, which can be traced both on the dependences of the configurational energy, long-range and short-range order parameters, and on changes in the atomic structure and distributions of structural-phase states. Both ordered and disordered phases, a certain set of superstructural domains, are realized simultaneously. This means that in the vicinity of the disorder–order phase transition low-stability states are observed.

It is shown that the influence of planar defects (antiphase boundaries (APBs)) on the disordering process is significant up to the temperature of the structural-phase transformation. The most significant for the long-range order is the appearance of the defect itself; the difference in the type of antiphase and their plane of occurrence does not affect the behaviour of the long-range order with temperature. The type of an APB has a noticeable effect on the structural and energy characteristics of the system at temperatures below the phase transformation temperature. Logically, a system with structural defects is less ordered than a defect-free system. The presence of a defect contributes to the onset of disordering of the system at lower temperatures: a decrease in the order in the alloy begins in the case of thermal APBs at a lower temperature compared with the case of shear APBs. The presence of antiphase boundaries affects the stability of the alloy upon heating. The disordering process is accompanied by blurring of the boundaries and their faceting.

From a comparative analysis of the features of disordering processes in the BCC system (the traditional CuZn alloy and the NiAl intermetallic compound) with increasing temperature in the region of low-stability pre-transitional states, it follows that if the order–disorder phase transition occurs in the traditional CuZn alloy as a result of disordering in the system, then in the NiAl intermetallic compound long-range order occurs as a result of structural-phase transformation. In fact, in the traditional CuZn alloy, the disordering process dominates, and in the NiAl intermetallic compound, the structural-phase transformation is dominant.

Metal materials with a BCC structure are currently widely used because of their unique physical and mechanical properties. Two groups can be distinguished among them, whose representatives have excellent structural and phase features and behavioural patterns, especially in the high-temperature region of low-stability pre-transitional states: traditional alloys (for example, alloys of the Cu–Zn system) and intermetallics (for example, alloys of the Ni–Al system).

Copper and its alloys are traditionally widely used due to the combination of high mechanical and technological properties. Compared to copper, brass has higher strength, corrosion resistance, better casting properties, and a higher recrystallization temperature. Copper and zinc form both α solid solutions based on copper, and a number of intermediate phases β, γ, etc. Phase β is a solid solution with a BCC lattice based on the compound CuZn (Hume-Rothery phase). Upon cooling, the β-phase at a temperature of ~450°C changes to an ordered state (β → β'), and the β'-phase is harder and more brittle than the β phase. Phase γ is a solid solution with a BCC lattice based on the Cu_5Zn_8 compound, which is very brittle, therefore, its presence in industrial alloys is undesirable. Naturally, the mechanical properties of brass are determined by the properties of the phases, and with increasing zinc content, their strength increases. The maximum strength is achieved in the two-phase region (α + β) with a zinc content of about 45%, therefore, α and (α + β) brass are mainly used.

A characteristic feature of alloys of the Ni–Al system is a high ordering energy. The NiAl intermetallic compound and solid substitutional solutions based on it have a high degree of long-range order, which remains in the entire temperature-concentration region of their existence up to the melting point. The intermetallic monoaluminide NiAl has a BCC lattice ordered by type $B2$, in which two simple cubic sublattices of nickel and aluminum can be distinguished. The Ni–Al system is characterized by a large difference in atomic sizes and electronic structure.

The presence of mixed covalent, ionic, and metal interatomic bonds in NiAl determines a large unit cell volume and a large Burgers vector, a decrease in independent equivalent slip systems, a complexity of the reactions of dislocation interactions with each other, with various boundaries and stacking faults, determines slip localization, and makes it difficult to transfer strain across the boundaries. The large magnitude of the interatomic interaction forces in the lattice of nickel monoaluminide also determines mainly the properties of β alloys [1].

Classical studies of low-stability pre-transitional states have been performed on β-alloys of the Ni–Al system [1]. An important direction is the study of competition and the mutual influence of parallel processes (ordering and decay of a β-solid solution, ordering and microdisintegration, ordering and martensitic transformation) and the regulation of complex atomic ordering processes in order to increase the structural stability and mechanical properties of heat-resistant β-phase-based intermetallic compounds of the Ni–Al systems. It is known that nickel monoaluminide is characterized by a high melting point (1638°C) and a high heat of formation. NiAl crystals exhibit strong elastic anisotropy and related anisotropy of properties unlike the structures with a disordered BCC lattice [1].

Naturally, the properties of alloys are associated with the structural-phase state of materials, and ultimately with the properties and structure of phases, which commonly have structural defects. The study of the properties and structural-phase state of materials by computer simulation methods allows us to study in detail the mechanisms of the ongoing physical and chemical processes [2–4]. Antiphase boundaries are a special type of plane defects. A characteristic feature of a shear APBs is that all atoms located on one side of the boundary plane are shifted by a vector connecting the atoms of two sublattices relative to atoms on the other side of the boundary. For the $B2$ superstructure, such a shift corresponds to a change in the sorts of all atoms by the sorts of the opposite component. In alloys with a $B2$ superstructure, shear APBs (SAPBs) are formed in planes with an even sum of Miller indices, and thermal (TAPBs) with an odd one. TAPBs in the $B2$ superstructure are formed predominantly in the planes of the cube and octahedron. An important characteristic of any APB is the energy of its formation. The lower the APB formation energy – surface tension – the greater the distance the dislocations diverge.

The properties of alloys are associated with the structural-phase state of materials, ultimately, with the properties and structure of the phases making up the alloy. The study of the properties and structural-phase states of materials by experimental methods is laborious, and often the mechanisms of the occurring physicochemical processes are very difficult to identify. The situation is complicated by thermal power loading. Therefore, systematic studies of the structural-phase states of metal systems by computer simulation methods attract close attention, since it is possible to reveal the physicochemical processes and phenomena occurring in the system [1–3]. The most common method at present is the molecular dynamics method; most studies of intermetallic compounds of the Ni – Al system and brass are carried out by this method. For example, the features of low-stability structural-phase states in FCC Cu–Pt systems and in the BCC system were examined using the example of the Ni–Al system [4–15].

1.1. Approximations and the model used

Consider an ordered BCC structure with a $B2$ superstructure (CuZn alloy) (Fig. 1.1). The computational grid includes $32 \times 32 \times 32$ unit cells (65536 atoms), and we use periodic boundary conditions.

The interaction between different pairs of atoms of the alloy components will be specified using the semi-empirical Morse pair potential in the form of a function

$$\varphi(r_{ij}) = D_{KL}\beta_{KL}e^{-\alpha_{KL}r_{ij}}\left(\beta_{KL}e^{-\alpha_{KL}r_{ij}} - 2\right),$$

where α_{KL}, β_{KL} and D_{KL} are the parameters of the potentials describing the bonds of pairs of atoms of varieties $K–L$; r_{ij} is the distance between the atoms. The configurational energy of the crystal will be calculated as

$$E = 1/2\sum_{i=1}^{N}\sum_{j=1}^{M}\varphi(r_i - r_j),$$

where $r_i–r_j$ is the distance between atoms i and j; N is the number of atoms in the crystal; M is the number of nearest neighbours.

In the study, we will pay special attention to changes in configurational energy, short-range and long-range order parameters in two parts of each thermal cycle: step heating from 200 to 1600 K and subsequent step cooling to the initial temperature.

For calculations, we use the Metropolis Monte Carlo Method. To activate the diffusion process, one vacancy is randomly introduced into the system, which corresponds to a concentration of ~$1.81\cdot10^{-5}$. We use only the vacancy diffusion mechanism. The state of the alloy will change at discrete points in time, for one iteration we take one act of self-diffusion, corresponding to the jump of an atom to a vacant site. The dynamic or kinetic component is present only in jumps of atoms to vacant sites. We assume that the state of the system can change only at discrete time instants with a step Δt. In this work, the transition to real time is not carried out; therefore, the duration of each experiment is determined in arbitrary units of time equal to the number of jumps of atoms to the vacant sites, i.e. $\Delta t = 1$ corresponds to one iteration. At each iteration, the probability

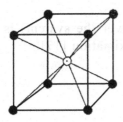

Fig. 1.1. Unit cell of $B2$ superstructure.

of hopping the atom i closest to the vacant site, located at a distance of up to three coordination spheres from it, to the position of this vacancy is calculated as follows:

$$p_i = A^{-1}e^{-\frac{E_{\max}-\left(E^i-E^i\right)}{kT}},$$

where A^{-1} is the energy of the i-th atom in the initial position; A^i_e is the energy of the i-th atom after jumping to the position of a vacancy; A is the normalization constant; $E_{\max}$ is the maximum difference between the initial and final atomic energies $E_{\max} = \max_{0<i<M}\left(E^i_i-E^i_f\right)$. To determine the normalization constant A, the following partition of the segment was constructed:

$$0 = A_0 < A_1 < A_2 < A_3 < ... < A_{M-1} < A_M = 1$$

so that $|A_i-A_{i-1}|$. Then, using a random number generator, a number was selected and the segment of the partition to which this number belonged was determined, i.e. there was an atom with index j from the condition.

For each temperature, 106 iterations were performed, and the temperature change step was 100 K.

In the study, special attention is paid to changes in configurational energy and short-range and long-range order parameters, as well as structurally-phase low-stability states in the process of order-to-disorder transformation.

The short-range order parameter on the i-th sphere is defined in the Cowley approximation [16]:

$$\sigma^{AB}_i = 1 - \frac{P^{AB}_i}{C_B}$$

where $\tilde{N}_{\tilde{A}}$ is the concentration of atoms of component B; P^{AB}_i is the probability of the formation of the A–B bond for the i-th atom in the ith coordination sphere.

The long-range order parameter (averaged over the system) is calculated in the Gorsky–Bragg–Williams approximation [17]:

$$\eta = \frac{P^{(1)}_A-C_A}{1-\nu},$$

where $P_A^{(1)}$ is the probability of filling with atoms of component A of nodes of the first type; C_A the concentration of atoms of component A in the alloy; v is the concentration of nodes of the first type.

1.2. Features of structural-phase transformations of the CuZn alloy during the first thermal cycle

Let us consider the structural features and energy characteristics of the BCC CuZn alloy in the course of the first thermal cycle, i.e. in the process of the order–disorder and disorder–order phase transitions. For calculations, we use the Metropolis Monte Carlo Method. In describing the interatomic interaction, we use the parameters of Morse potentials given in Table 1.1. The potential values were tabulated as changes in energy depending on interatomic distances.

Table 1.1. Morse potential parameters for CuZn alloy

Type of interaction	α, Å^{-1}	β	D, eV
Cu–Zn	1.495109	41.598	0.3736
Cu–Zn	1.447832	35.607	0.322
Zn–Zn	1.71223	81.104	0.2189

Dependences of the configurational energy and short-range order parameter on the first coordination sphere on temperature during the order–disorder–order phase transitions are shown in Fig. 1.2.

It is easy to see that during thermal cycling in the region of low-stability states of the system, a hysteresis loop is observed. This indicates that during heating and cooling the system undergoes structural-phase states that differ from each other.

According to the graphs, significant differences in the structural and energy characteristics of the alloy during phase transitions of the order–disorder and disorder–order transitions are observed in the temperature range from 600 to 1200 K. For temperatures below 600 K, the energy practically does not change, but differs from the corresponding values for the initial alloy configurations and alloy configurations after the disorder–order phase transition. The difference in energy values indicates that the characteristics of the alloy after completion of the heating-cooling cycle differ from those of the model alloy corresponding to the ordered $B2$ superstructure.

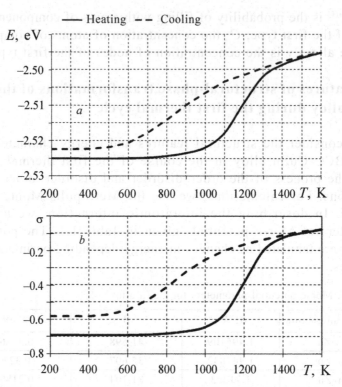

Fig. 1.2. Hysteresis loops in the change in configurational energy (*a*) and short-range order (*b*) in the CuZn alloy during thermal cycling.

From the dependences of the long-range order parameter in the process of thermal cycling (Fig. 1.3), it is easy to see that during the order–disorder phase transition with a decrease in temperature to $T = 700$ K there is no violation of the long-range order. In the temperature range 700–1100 K, a gradual decrease in its values is seen, and upon further heating, a sharp disorder in the vicinity of $T \approx 1200$ K. With a further increase in temperature, the long-range order tends to zero, which indicates the absence of a long-range order in the alloy. In the process of the disorder–order phase transition, the long-range order is absent up to $T \approx 800$ K.

When the alloy is cooled during the order–disorder phase transition (Fig. 1.3), no long-range order is observed when the temperature drops to $T \approx 800$ K, although the short-range order begins to appear much earlier (Fig. 1.2 *b*). A sharp increase in the values of the long-range order parameter is observed when the temperature decreases to about 600 K. With further cooling, its

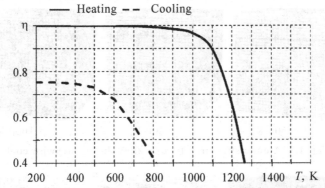

Fig. 1.3. Temperature dependence of the long-range order parameter during thermal cycling.

Fig. 1.4. Atomic structure of CuZn alloy during step heating.

values gradually increase, but do not reach unity. The temperature range of variation of the long-range order parameter (Fig. 1.3) is consistent with that of configurational energy (Fig. 1.2 *a*).

Let us consider the change in the atomic structure of the alloy as a function of temperature during order–disorder phase transitions and the disorder–order transitions. In the process of stepwise heating (Fig. 1.4) to a temperature of ~600 K, the alloy is ordered in accordance with the *B*2 superstructure.

At $T \approx 800$ K, the first disordered regions appear, and with a further increase in temperature to about 1200 K the number and size of regions with a deviation from the superstructural arrangement increase. At temperatures above 1400 K there is no superstructural ordering.

In the process of stepwise cooling (Fig. 1.5) to a temperature of ~1200 K, the alloy is completely disordered; at ~1000 K, the first regions ordered in accordance with the *B*2 superstructure appear. With further cooling, the number and size of ordered regions increases, and at $T \approx 800$ K, the formation of two ordered domains is noticeable.

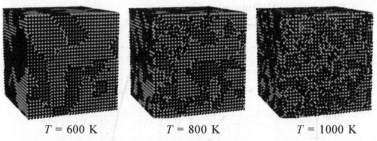

T = 600 K   T = 800 K   T = 1000 K

Fig. 1.5. Atomic structure of CuZn alloy during stepwise cooling.

Fig. 1.6. Distributions of atoms over ordered and disordered phases during heating.

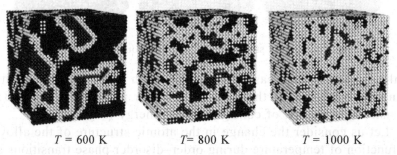

T = 600 K   T= 800 K   T = 1000 K

Fig. 1.7. Distributions of atoms over ordered and disordered phases during cooling.

The domain sizes are increasing, the boundaries are becoming clearer for temperatures below 600 K.

The change in the domain structure of the CuZn alloy during the order–disorder and disorder–order phase transitions is presented in Figs. 1.6 and 1.7, respectively. Domains are marked in dark, and unordered domains and domain boundaries are highlighted in light.

It is easy to see that during heating to $T = 400$ K the alloy is completely ordered and has a single-domain structure. The first disordered regions begin to appear in the temperature range of 600–800 K, and an increase in temperature to 1200 K leads to an increase in the number and size of disordered regions; small-sized domains of the $B2$ superstructure are observed. With a further increase in temperature to 1400 K and higher, almost the entire alloy is disordered, only domain nuclei remain.

During cooling to 1400 K, only the $B2$ superstructure nuclei are observed in the alloy, the sizes of which increase, for example, at 1000 K. Cooling to 800 K leads to a gradual increase and merging of the of ordered phases. For a lower temperature (~600 K), no disordered regions remain in the alloy except the domain boundaries. With a further decrease in temperature, the shapes of the boundaries change, domains of the same type join and remain so up to 200 K. After the disorder–order phase transition is complete, two domains of the $B2$ superstructure are formed in the system.

Comparing the features of the behaviour of the configurational energy (see Fig. 1.2 a) during heating, the long-range (see Fig. 1.3) and short-range (see Fig. 1.2 b) order parameters, the change in the atomic structure of the alloy during the order–disorder phase transition is disorder (see Fig. 1.4), and the distribution of the atoms over ordered and disordered phases during heating (Fig. 1.6), we can conclude that the structural-phase transformation occurs in a certain temperature range $T \approx 1100–1200$ K. In this range, there are very small thermodynamic preferences of one or another structural state that is easy to follow on both the dependences of the configuration energy parameters of the long and short-range order and also the atomic structure changes and the distribution of structural and phase states. The simultaneous coexistence of ordered and disordered phases is observed, as well as a set of superstructural domains. This indicates the presence of precisely low-stability states in the vicinity of the order–disorder phase transition.

From such a comparison (see Figs. 1.2, 1.3, 1.5, and 1.7), we can conclude that during cooling the structural-phase transformation occurs already in a different temperature range $T \approx 700–800$ K. In this interval, the thermodynamic stimuli of one or the other structural-phase state are very small, which can also be traced both on the curves of dependence of configurational energy, long-range and short-range order parameters, and on changes in the atomic structure and distributions of structural-phase states. Both ordered and disordered phases, a certain set of superstructural domains, are realized simultaneously. This means that low-stability states are realized in the vicinity of the order–disorder phase transition.

A hysteresis loop is observed during thermal cycling in the region of low-stability states of the system: when heated and cooled, the system goes through structural-phase states that differ from each other. Of particular note is the domain structure in the region of low-stability states: the 'entangled' domain structure is characteristic of the interval of low-stability states, which indicates the absence of thermodynamic preferences of certain phases or domains. The 'entangled' domain structure indicates precisely the thermodynamically weakly defined structural state of the system, in which different structural components of the system may differ in the sense of symmetry. From the form of low-stability states, it is not difficult to make a conclusion on a simultaneous coexistence of both parent and daughter structures during the order–disorder phase transition in these low-stability states. The thermodynamic stimuli of transformation are very small; therefore, a variety of possible structural states can be observed in the system.

Thus, during thermal cycling, the irreversibility of states in the process of order–disorder and disorder–order phase transitions should be noted. Moreover, in the vicinity of the order–disorder and disorder–order phase transitions low-stability structural-phase states of the system are realized, with some set of them constantly changing depending on temperature. It can be assumed that when the alloy is heated (during the order–disorder phase transition), overheating of the alloy occurs relative to the traditionally understood order–disorder phase transition temperature. During cooling (during the disorder–order transition), supercooling occurs. This temperature difference is naturally related to the width of the hysteresis loop. It can be assumed that during repeated heating-cooling cycles, the loop width will narrow, which will affect the decrease in the temperature difference between the order–disorder and disorder–order phase transitions.

Conclusion. Using the Monte Carlo method, the structural and energy characteristics of β-brass have been studied during thermal cycling. As a result of the heating and cooling cycle, a kind of hysteresis is observed, the presence of which indicates the irreversibility of the processes that take place, which implies a difference in the structural-phase states at the heating and cooling stages. Analysis of the atomic and phase structure of the system during heating and cooling, i.e. in the process of order–disorder and disorder–order phase transitions, showed that the system goes through different structural-phase states. After the completion of the disorder–order phase transition in the system, two domains of the B2 superstructure are formed.

Using the Monte Carlo method, it has been shown that irreversibility of processes is observed during thermal cycling in the course of structural-phase transformations in the CuZn alloy. Moreover, in the vicinity of the order–disorder and disorder–order phase transitions low-stability structural-phase states of the system appear, with some of them constantly changing depending on temperature.

During thermal cycling in the region of low-stability states of the system, a hysteresis loop is observed: when heating and cooling, the system goes through structural-phase states that differ from each other. Of particular note is the domain structure in the region of low-stability states: the 'entangled' domain structure is characteristic of the interval of low-stability states, which indicates the absence of thermodynamic preferences of certain phases or domains. The 'entangled' domain structure indicates precisely the thermodynamically weakly defined structural state of the system, in which different structural components of the system may differ in the sense of symmetry. From the form of low-stability states, it is not difficult to conclude that during the order–disorder phase transitions in these low-stability states both parent and daughter structures coexist simultaneously. The thermodynamic stimuli of transformation are very small; therefore, a whole spectrum of possible structural states can be observed in the system.

Thus, using the Monte Carlo method, the irreversibility of structural-phase transitions during thermal cycling against the background of the combination of atomic ordering and structural rearrangements in a system has been demonstrated.

1.3. Structural-phase transformations of the CuZn alloy during thermal power cycling

We studied by the Monte Carlo method the structural features and energy characteristics of the BCC CuZn alloy during several thermal cycles (successive heating-cooling cycles, i.e. successive order–disorder and disorder–order phase transitions).

In describing the interatomic interaction, we use the parameters of the Morse potentials given in Table 1.1. The potential values were tabulated as changes in energy depending on interatomic distances.

The dependence of the configuration energy on temperature during several successive order–disorder and disorder–order phase transitions during several successive heating-cooling cycles is shown in Fig.

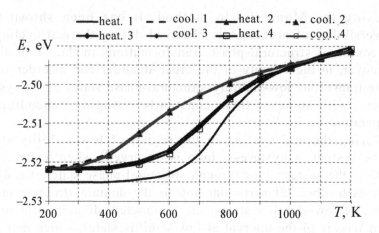

Fig. 1.8. Temperature dependence of the configurational energy during several successive order–disorder and disorder–order phase transitions in the course of a number of successive heating-cooling cycles (hysteresis loops in the change in configurational energy during thermal cycling).

1.8. On the curve of dependence of the configurational energy, it is easy to see that during thermal cycling in the region of low-stability states of the system, a hysteresis loop is observed. This indicates that during heating and cooling the system undergoes structural-phase states that differ from each other. It should be noted that the hysteresis loop closes starting from the second cycle. At the stage of heating the second cycle, the system does not repeat the thermodynamic states of the stage of heating the first cycle. It is easy to see (Fig. 1.8) that in both the first and subsequent heating-cooling cycles, significant differences in the energy characteristics of the alloy during the order–disorder and disorder–order transitions are observed in the temperature range from 0 to ~800 K. For temperatures below 400 K, there are practically no changes, however, the energy differs from the corresponding values for the initial alloy configuration and the system configuration after the disorder–order phase transition. The difference in the energy values at the end of the first cycle indicates that the characteristics of the alloy after completion of the heating-cooling cycle differ from the characteristics of the model alloy corresponding to the ordered $B2$ superstructure. The thermodynamic states during thermal cycling during the heating and cooling processes in the second cycle differ from the corresponding states of the first cycle, the hysteresis loop closes starting from the second cycle. It should be noted that during

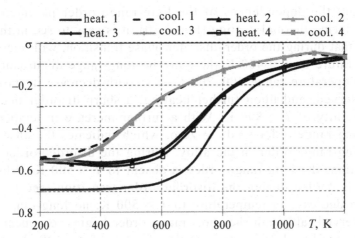

Fig. 1.9. Hysteresis loops in the variation of the short-range order parameter change during thermal cycling.

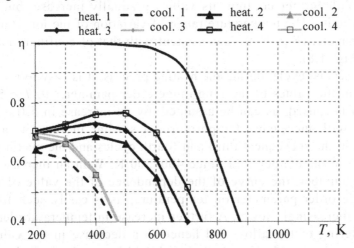

Fig. 1.10. Hysteresis loops in the variation of the long-range order parameter η.

repeated cycles of thermal cycling, the loop width decreases (Fig. 1.8), as it does in experimental studies of the thermal-force cycling [11]. However, in the second and subsequent cycles, no differences are observed, which is associated with the features of the applied computational algorithm.

Consider the behaviour of the integral structural characteristics during thermal cycling: a short-range order parameter (Fig. 1.9) and a long-range order parameter (Fig. 1.10). The temperature dependence of the short-range order parameter on the first coordination sphere [16] (Fig. 1.9) repeats the shape of the configuration energy curve (Fig. 1.8).

From the dependences of the long-range order parameter $\eta = \eta(T)$ during thermal cycling (Fig. 1.10), it is easy to see in the first cycle that during the order–disorder phase transition, with increasing temperature to $T = 400$ K there are no long-range disturbances. In the temperature range 400–700 K, a smooth decrease in its values is observed, and upon further heating, a sharp disorder occurs in the vicinity of $T \approx 800$ K. With a further increase in temperature, the long-range order tends to zero, which indicates the absence of the long-range order in the alloy. During the disorder–order phase transition, the long-range order is absent to $T \approx 500$ K. When the alloy is cooled during the disorder–order transition (see Fig. 1.2) with a decrease in temperature to $T \approx 500$ K, no long-range order is observed, although the short-range order begins to appear much earlier (see Fig. 1.3). A sharp increase in the values of the long-range order parameter is observed when the temperature decreases to about 400 K. With further cooling, its values gradually increase, but do not reach unity. The temperature range of variation of the long-range order parameter (Fig. 1.10) is consistent with that of configurational energy (Fig. 1.8).

In the second cycle, during the heating process, it is easy to see an increase in the values of the long-range order parameter to $T \approx 500$ K, which is caused, as can be assumed, by an increase in diffusion mobility with increasing temperature. A similar increase is also observed in the subsequent third and fourth cycles and the maximum values lie in the temperature range 400–500 K. With an increase in the cycle number, the curve of the dependence of the value of the long-range order parameter on temperature, as is easily seen from Fig. 1.9, increasingly lower. A further increase in temperature leads to disordering of the alloy, and hence to a decrease in the values of the long-range order parameter. It should be emphasized that as the cycle number increases, the temperature at which the system is completely disordered decreases.

During the cooling process in the second cycle, it can be seen (Fig. 1.10) that the appearance of non-zero values of the long-range order parameter is observed at the same temperatures as in the first cycle. Moreover, the dependence curve $\eta = \eta(T)$ with decreasing temperature lies below the corresponding curve of the first cycle. At $T \to 0$, the values of the long-range order parameter do not reach the corresponding values of the first cycle. The value of the long-range order parameter at $T \to 0$ decreases with increasing cycle number. With further cycling (increasing the cycle number n), the dependence curves $\eta = \eta(T)$ with decreasing temperature lies increasingly lower as the cycle number increases.

The behaviour of the long-range order parameter and the short-range order parameter during thermal cycling indicates that the structural-phase states of the system do not repeat when the system is either heated or cooled. As in the process of increasing the temperature the transition is an order–disorder transition, and in the process of lowering the temperature, the transition is a disorder–order transition, the system passes through different structural-phase states. As the cycle number increases, these differences decrease, as evidenced by the approximation of the curves of the dependences of the long-range order parameter in the process of thermal cycling.

From the definition of free energy $F = E–T \cdot S \; (\eta(T))$, where E is the internal configurational energy of the system, T is the temperature on the Kelvin scale, $S(\eta)$ is the configurational entropy [17], and from the analysis (with increasing cycle number n) of the dependences $E = E(T)$ (see Fig. 1.2) and $S = S(\eta(T))$ based on the dependence $\eta = \eta(T)$ (Fig. 1.10) the following naturally follows. In the implementation of each cycle, the curve of dependence $F = F(T)$ has the form of a loop, the width of which decreases with increasing cycle number. The dependence realized in this way, i.e. a loop, is similar to that experimentally observed under thermal force loading during cycling [11].

Consider the change in the atomic structure of the alloy during the first cycle, depending on temperature during the order–disorder and disorder–order phase transitions. In the process of stepwise heating (Fig. 1.11 a) to a temperature of ~600 K, the alloy is ordered in accordance with the $B2$ superstructure. At $T \approx 400$ K, the first disordered regions appear, and with a further increase in temperature to about 700 K, the number and size of regions with a deviation from the superstructural arrangement increase. At temperatures above 1000 K, there is no superstructural order. In the process of stepwise cooling (Fig. 1.11 b) to a temperature of ~800 K, the alloy is completely disordered; at ~700 K, the first regions ordered in accordance with the $B2$ superstructure appear. With further cooling, the number and size of ordered regions increases, and at $T \approx 500$ K, the formation of two ordered domains is noticeable. The domain sizes are increasing, the boundaries are becoming clearer for temperatures below 400 K.

Of most interest is the change in the domain structure of the CuZn alloy during successive order–disorder and disorder–order

18

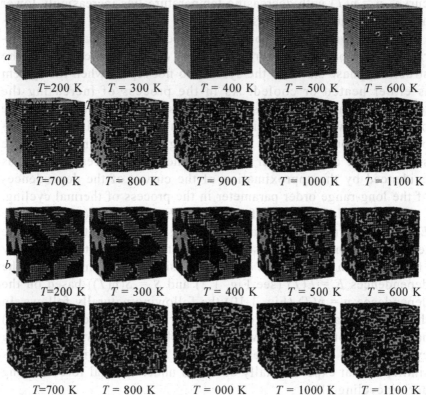

T=200 K   T = 300 K   T = 400 K   T = 500 K   T = 600 K

T=700 K   T = 800 K   T = 900 K   T = 1000 K   T = 1100 K

T=200 K   T = 300 K   T = 400 K   T = 500 K   T = 600 K

T=700 K   T = 800 K   T = 000 K   T = 1000 K   T = 1100 K

Fig. 1.11. Atomic structure of CuZn alloy during the first cycle step heating (*a*) and cooling (*b*).

transitions during several successive heating-cooling cycles. In Figs. 1.12–1.15, domains are marked in dark colour, and disordered regions and domain boundaries are marked in light. Let us analyze the change in the domain structure during the first heating-cooling cycle (Fig. 1.12).

It is easy to see that during heating (Fig. 1.12 *a*) to $T = 300$ K, the alloy is completely ordered and has a single-domain structure. The first disordered regions begin to appear in the temperature range 400–600 K, and an increase in temperature to 800 K leads to an increase in the number and size of disordered regions; small-sized domains of the $B2$ superstructure are observed.

With a further increase in temperature to 1100 K and higher, almost the entire alloy is disordered, only domain nuclei remain.

During cooling (Fig. 1.12 *b*) to 800 K, only nuclei of the $B2$ superstructure domains are observed in the alloy and their sizes increase (for example, at 700 K). Cooling to 500 K leads to a gradual increase and merging of the ordered phases. For a lower temperature (400 K), no disordered regions remain in the alloy except the domain boundaries. With a further decrease in temperature, the shape of the boundaries changes, domains of the same type join and are stored up to 200 K.

After the completion of the disorder–order phase transition in the system, two domains of the $B2$ superstructure are formed.

Comparing the features of the behaviour of the configurational energy during heating (see Fig. 1.2), the short-range (see Fig. 1.9) and long-range (Fig. 1.10) order parameters, the changes in the atomic structure of the alloy during the order–disorder phase transition (Fig. 1.11 a), the distribution of atoms over ordered and disordered phases during heating (Fig. 1.12 a), we can conclude that the structural-phase transformation occurs in a certain temperature range $T \approx 600$–800 K. In this range, the thermodynamic preferences of other structural-phase states are quite insignificant, which is easy

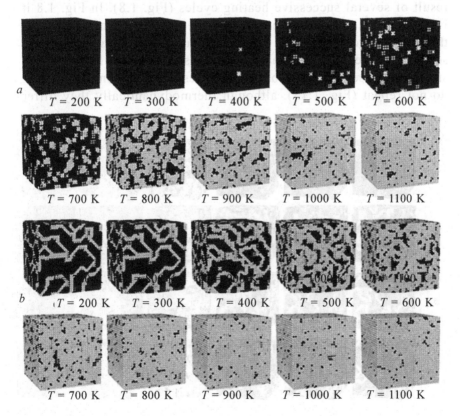

Fig. 1.12. Distributions of atoms over ordered and disordered phases during heating (a) and cooling (b).

to trace both on the dependences of the configurational energy, the long- and short-range parameters, and on the changes in the atomic structure and distributions of structural-phase states. A simultaneous coexistence of ordered and disordered phases and a set of superstructural domains are observed. This indicates the presence of precisely low-stability states in the vicinity of the order–disorder phase transition.

Let us pay attention to the change in the domain structure of the CuZn alloy during successive order–disorder and disorder–order phase transitions as a result of several successive heating-cooling cycles (Figs. 1.13 and 1.14).

It should be noted that in the first cycle during heating, a sequence of structures is observed (Fig. 1.13) from the corresponding sequences of other cycles. This was to be expected from the shape of the curve of dependence of the configurational energy on temperature during several successive order–disorder phase transitions as a result of several successive heating cycles (Fig. 1.8). In Fig. 1.8 it is easy to see that in subsequent cycles, starting from the second, the temperature dependence of the configurational energy on several sequential order–disorder phase transitions is repeated. However, the structural-phase states in subsequent cycles, starting from the second, do not repeat (Fig. 1.13), although thermodynamically they differ

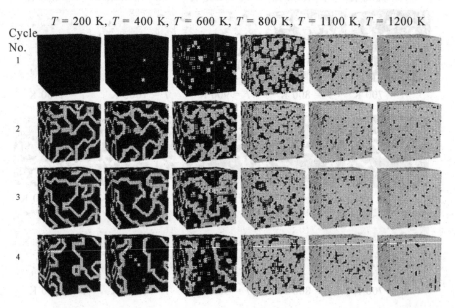

$T = 200$ K, $T = 400$ K, $T = 600$ K, $T = 800$ K, $T = 1100$ K, $T = 1200$ K

Cycle No.

Fig. 1.13. Distributions of atoms over ordered and disordered phases during heating.

Cycle No.

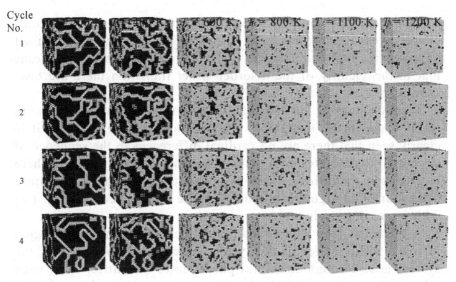

Fig. 1.14. Distributions of atoms over ordered and disordered phases during cooling

only slightly, as can be assumed from the temperature dependence of free energy in different cycles. The structural-phase states in the second and subsequent cycles differ; this is especially evident in the region of low-stability states of the system in a certain temperature range $T \approx 600–800$ K. The sequence of the images in Fig. 1.13 demonstrates that for two successive heating cycles at the same temperature, the structural-phase states differ. These differences are not dramatic, but occur within each cycle, and decrease as the cycle number increases. Actually, the system is training a tendency towards a certain steady sequence of structural-phase states. This could be expected from the behaviour of the dependences $\eta = \eta(T)$ with increasing n (Fig. 1.10).

Comparing the features of the behaviour of the configurational energy (see Fig. 1.8), the short-range (see Fig. 1.9) and long-range (Fig. 1.10) order parameters, the changes in the atomic structure of the alloy during the disorder–order transition in the course of cooling (Fig. 1.11 b), the distribution of atoms over ordered and disordered phases during cooling (Fig. 1.12 b), we can conclude that the structural-phase transformation occurs already in a different temperature range $T \approx 500–600$ K. In this range, the thermodynamic stimuli of realization of one or other structural-phase state are very small, which can also be seen on the curves dependence of the configurational energy, the short- and long-range order parameters and the changes in nuclear structure and distributions of structural-phase states. Both ordered and disordered phases and a certain set of superstructural domains, are realized simultaneously. This means that in the vicinity of the disorder–order transition there are low-stability states.

From changes in the distributions of structural-phase states with an increase in the number of the cooling cycle (Fig. 1.14) it is clearly seen that there are no cardinal differences in the sequence of structural-phase states with an increase in the cycle number. There is a certain evolution, which is consistent with the temperature behaviour of configurational energy in the cooling section (Fig. 1.8) with increasing cycle number.

A hysteresis loop is observed during thermal cycling in the region of low-stability states of the system: when heated and cooled, the system goes through structural-phase states that differ from each other. Of particular note is the domain structure in the region of low-stability states: the 'entangled' domain structure is characteristic of the interval of low-stability states, which indicates the absence of thermodynamic preferences of certain phases or domains. The 'entangled' domain structure indicates the thermodynamically weakly defined structural state of the system, in which different structural components of the system may differ in the sense of symmetry. From the form of low-stability states it is not difficult to conclude a coexistence of both parent and daughter structures during the order–disorder transition in these low-stability states. The thermodynamic stimuli of transformation are very small; therefore, a variety of possible structural states can be observed in the system.

Thus, during thermal cycling, the irreversibility of states in the course of order–disorder and disorder–order transitions should be noted. Moreover, in the vicinity of the order–disorder and disorder–order transitions there are low-stability structural-phase states of the system, with some of them constantly changing depending on temperature. It can be assumed that when the alloy is heated (during the order–disorder phase transition), an overheating of the alloy occurs relative to the traditionally understood order–disorder phase transition temperature. During cooling (during the disorder–order transitions), supercooling occurs.

This temperature difference is naturally related to the width of the hysteresis loop. It can be assumed that during repeated heating-cooling cycles the loop width will narrow, which will affect the decrease in the temperature difference of the order–disorder and disorder–order transitions and the convergence of structural-phase states (Fig. 1.15). It is easy to see in the dependence of the configurational energy that a hysteresis loop is observed during thermal cycling in the region of low-stability states of the system, and it closes during the second and subsequent cycles. This indicates that during heating and cooling the system undergoes structural-phase states that differ from each other. At the stage of heating in the second and subsequent cycles the system does not repeat the thermodynamic states of the stage of heating of the first cycle. It is easy to see (see Fig. 1.8) that there are significant differences in the energy characteristics of the alloy during the order–disorder and disorder–order phase transitions in both the first and subsequent heating-cooling cycles. The situation becomes even more expressive when considering the free energy $F = E-T \cdot S (\eta (T))$ with increasing cycle number n.

In this case, during the implementation of each cycle, the curve of dependence $F = F(T)$ has the shape of a loop, the width of which decreases with increasing cycle number. This, in turn, means that the evolution of the loop indicates a convergence of the structural-phase states with increasing cycle number, which is visually observed (Fig. 1.15) when the cycle number affects the distribution of atoms over ordered and disordered phases in low-stability states in heating (at $T = 700$ K) and cooling (at $T = 500$ K).

Thus, the research results demonstrate that for two successive heating-cooling cycles at the same temperature, the structural-phase states differ both at the heating stage and at the cooling stage. These differences are not revolutionary, but occur on each cycle, and decrease as the cycle number increases. Actually, the system

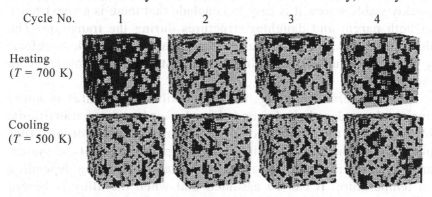

Fig. 1.15. Influence of the cycle number on the distribution of atoms in ordered and disordered phases from heating ($T = 700$ K) and cooling ($T = 500$ K).

is training with a tendency towards a certain steady sequence of structural-phase states.

Conclusion. The structural-phase transformations of β-brass in the process of thermal force cycling using the example of thermal cycling (several successive order–disorder and disorder–order phase transitions during several successive heating-cooling cycles) were studied using the Monte Carlo method. It is shown that a kind of hysteresis is observed as a result of each cycle of heating and cooling and its presence indicates the irreversibility of the processes that occur, which implies a difference in the structural-phase states both at the heating and cooling stages. Analysis of the atomic and phase structure of the system during heating and cooling, i.e. in the process of order–disorder and disorder–order transitions, has shown that the system goes through different structural-phase states.

In this case, in the vicinity of order–disorder and disorder–order phase transitions there are low-stability structural-phase states of the system with a set of phase and superstructural domains that are constantly changing depending on temperature. A hysteresis loop is observed during thermal cycling in the region of low-stability states of the system: when heating and cooling, the system goes through structural-phase states that differ from each other. Of particular note is the domain structure in the region of low-stability states: the 'entangled' domain structure is characteristic of the range of low-stability states, which indicates the absence of thermodynamic preferences of certain phases or domains. The 'entangled' domain structure indicates the thermodynamically weakly defined structural state of the system, in which different structural components of the system may differ in the sense of symmetry. From the type of weakly stable states, it is easy to conclude that there is a coexistence of both parent and daughter structures during the transition. The thermodynamic stimuli of transformation are very small; therefore, a whole spectrum of possible structural states can be observed in the system.

During thermal cycling, the irreversibility of states is noted during the order–disorder and disorder–order phase transitions. Moreover, in the vicinity of order–disorder and disorder–order phase transitions, low-stability structural-phase states of the system are realized with some set of them constantly changing depending on temperature. It can be assumed that when the alloy is heated (during the order–disorder phase transition), overheating of the alloy occurs relative to the traditionally understood order–disorder phase transition temperature. Supercooling occurs during cooling (during the disorder–order phase transition process).

From a comparison of the specific features of the behaviour of configurational energy, short-range and long-range order parameters, changes in the atomic structure of the alloy, distribution of atoms in ordered and disordered phases, it is concluded that structural-phase transformations at the stages of heating and cooling occur in different temperature ranges. In these ranges, the thermodynamic incentives for the realization of a structural-phase state are very small, which can be traced on the curves of the dependences of configurational energy, long-range and short-range order parameters, changes in the atomic structure, and distributions of structural-phase states. Both ordered and disordered phases and a certain set of superstructural domains are realized simultaneously. This means that low-stability states are observed in the vicinity of the disorder–order phase transition.

The research results demonstrate that the structural-phase states for two successive heating-cooling cycles at the same temperature differ both at the heating and cooling stages. These differences are not dramatic, but occur on each cycle, and decrease as the cycle number increases. The system is being trained with a tendency towards a certain steady sequence of structural-phase states.

1.4. Structural-phase features of the order-disorder phase transition in a BCC alloy with *B*2 superstructure in the presence of a complex of thermal antiphase boundaries

Computer simulation of plane defects in ordered alloys is relevant in connection with the need to explain and, if possible, control the physico-mechanical properties of alloys. Antiphase boundaries (APBs) are a special type of plane defects characteristic only for ordered alloys. The main feature of APBs is that all atoms located on one side of the boundary plane are shifted by a vector connecting the atoms of two sublattices relative to atoms on the other side of the boundary.

For the *B*2 superstructure, such a shift corresponds to a change in the sorts of all atoms. In alloys with *B*2 superstructure, shear APBs are formed in planes with an even sum of Miller indices, and thermal APBs (TAPBs) with odd ones. TAPBs in the *B*2 superstructure are formed mainly in the planes of the cube and octahedron [3]. It is known [18–23] that in alloys similar to the one under consideration, during the order–disorder transition, antiphase boundaries are smeared both interms of both positions of atoms in the vicinity of the boundaries [2–4] and in the atomic composition of the boundary regions [18–23]. The features of structural-phase states in the FCC Cu–Pt systems; and in the BCC system were exemplified by NiAl [7–10] and CuZn [11–15].

Using the Monte Carlo method, we study the effect of the complex of thermal antiphase boundaries in the <100> direction on the low-stability structural-phase states of β-brass (using the CuZn alloy as an example) during an order–disorder phase transition depending on the distance between the boundaries.

First, we consider a defect-free system (a defect-free CuZn model alloy) in the course of several order–disorder and disorder–order successive phase transitions during several successive heating-cooling cycles (we will call this process thermal cycling). In the future, this defect-free system will act as the initial state.

TAPBs normal to the <100> direction are considered. In this direction, the planes of nodes that are appropriate for Cu and Zn atoms alternate in the $B2$ superstructure (see Fig. 1.1). Thermal antiphase boundaries are specified by subtracting the planes of Cu or Zn atoms. Subtracting the plane of the Cu atoms, we obtain a thermal antiphase boundary and pairs of the nearest neighbours Zn–Zn (hereinafter, we will call this TAPB a boundary of the Zn–Zn type). Subtracting the plane of the Zn atoms, we obtain the thermal antiphase boundary and the pairs of nearest Cu–Cu neighbours.

hereinafter, we will call such a TAPB a Cu – Cu-type boundary). The Zn–Zn and Cu–Cu type boundary make up the dual complex in which the thermal antiphase boundaries are spaced apart by a certain distance. Note that with the introduction of such a dual complex, the equiatomic composition of the system does not change. In the presented images of the structural-phase state of the system, the Zn – Zn type boundary will be located on the right, and the Cu – Cu type boundary will be located on the left.

We draw attention to changes in the low-stability structural-phase state of the system as a result of the action of the vacancy diffusion mechanism during the order-disorder phase transition in a system with a pair of TAPBs (dual complex) separated by several unit cells in the <100> direction.

When finding the long-range order parameter

$$\eta = \frac{P_A^{(1)} - C_A}{1 - \nu}$$

it is necessary to calculate the probability $P_A^{(1)}$. To do this, we determine the number of atoms of sort A, in which the neighbours on the first sphere correspond to the first and second type of domains (parts of the system located on opposite sides of the antiphase boundary). Accordingly, nodes of the first type are considered to be all nodes that are legal for atoms of type A, depending on the type of domain:

$$P_A^{(1)} = \frac{N' + N''}{N_1},$$

where N_1 is the number of nodes of the first type; N' is the number of atoms of type A in the nodes of the sublattice of the first type; N'' is the number of atoms of sort A located in the nodes of another sublattice and ordered in accordance with the domain of the second type on the first sphere.

In describing the interatomic interaction, we use the parameters of Morse potentials given in Table 1.1. The potential values were tabulated as changes in energy depending on interatomic distances.

The dependence of the configurational energy on temperature during several successive order–disorder and disorder–phase phase transitions during several successive heating-cooling cycles is shown in Fig. 1.16 a. On the dependence of the configurational energy, it is easy to see that a hysteresis loop is observed during thermal cycling in the region of low-stability states of the system. This indicates that during heating and cooling the system undergoes structural-phase states that differ from each other. It should be noted that the hysteresis loop closes starting from the second cycle. At the stage of heating the second cycle, the system does not repeat the thermodynamic states of the stage of heating in the first cycle. It is easy to see (Fig. 1.16 a) that both in the first and subsequent heating-cooling cycles, significant differences in the energy characteristics of the alloy during order–disorder and disorder–order phase transitions are observed in the temperature range from about 400 to 800 K. For temperatures below 400 K, there are practically no changes, but the energy differs from the corresponding values for the initial configuration of the alloy and the configuration of the system after the disorder–order phase transition. The difference in the energy values at the end of the first cycle indicates that the characteristics of the alloy after the completion of the heating-cooling cycle differ from the characteristics of the model alloy corresponding to the ordered $B2$ superstructure. Thermodynamic states during thermal cycling in the heating and cooling processes in the second cycle

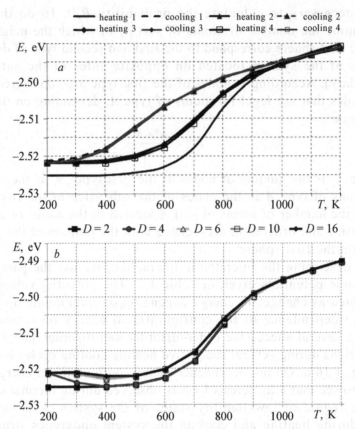

Fig. 1.16. Dependence of configuration energy on temperature: *a* – during several successive order–disorder and disorder–order phase transitions, during several successive heating-cooling cycles (hysteresis loops in a change in configuration energy during thermal cycling); *b* – during the order–disorder and disorder–order phase transitions in a system with a pair of TAPBs (dual complex) separated by 2, 4, 6, 10, and 16 unit cells in the <100> direction.

differ from the corresponding states of the first cycle, the hysteresis loop closes starting from the second cycle. It should be noted that during repeated cycles of thermal cycling, the loop width decreases (Fig. 1.16 *a*), as it does in experimental studies of the thermal force cycling [3, 13–15]. However, in the second and subsequent cycles, no differences are observed, which is associated with the features of the applied computational algorithm.

From the above analysis it follows that the region of low-stability states during structural-phase transformations lies in the temperature range of ~500–800 K.

Consider the temperature dependence of the configurational energy of the system at various distances between thermal antiphase boundaries (Fig. 1.16 *b*).

The temperature dependence of the configurational energy during the order–disorder phase transition in a system with a pair of TAPBs (Fig. 1.16 *b*) separated by 2 unit cells in the <100> direction (D = 2) is similar to the corresponding temperature dependence of the order–disorder phase transition in the first cycle in a defect-free system (Fig. 1.16 *a*). This leads to the conclusion that diffusion

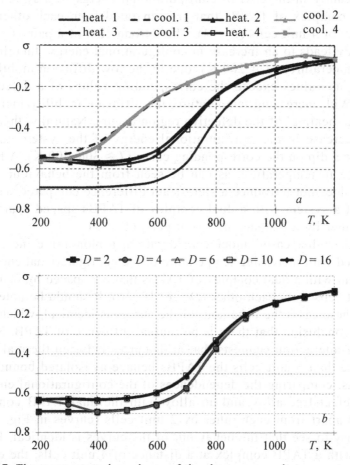

Fig. 1.17. The temperature dependence of the short-range order parameter on the first coordination sphere: *a* – during several successive order–disorder and disorder–order phase transitions during several successive heating-cooling cycles (hysteresis loops in the change in configurational energy during thermal cycling); *b* – during the order –disorder transition in a system with a pair of TAPBs (dual complex) separated by 2, 4, 6, 10, and 16 unit cells in the <100> direction.

processes during the order–disorder phase transition in a system with a pair of thermal antiphase boundaries located at a distance of 2 unit cells transfer the system to a defect-free state. It can be expected that diffusion due to the strong interaction of thermal antiphase boundaries 'erodes' the structural-phase disturbances introduced by a pair of 2 TAPBs separated by two unit cells. This can be attributed to the significant interaction of a pair of Zn–Zn-type and Cu–Cu-type boundaries at distance.

The shape of the curve of the temperature dependence changes significantly in the case of configurational energy of a system with a dual complex of TAPBs that are separated from each other by a distance of 4 unit cells ($D = 4$). The introduction of a pair of TAPBs naturally leads to an increase in configurational energy; therefore, at low temperatures ($T \approx 200$ K), due to the low diffusion mobility of atoms, its values exceed the corresponding values of a defect-free alloy. With increasing temperature, the diffusion mobility increases, and the 'healing' of the defective region occurs. Naturally, this leads to a decrease in the configurational energy of the system, and we observe a dip on the corresponding curve ($T \approx 300$–500 K). A further increase in temperature, as can be seen from the behaviour of the curve, does not introduce a significant difference compared with the case of a system with a dual complex of TAPBs spaced apart from each other by a distance of 2 unit cells ($D = 2$).

In the other cases under consideration, a monotonic increase is observed with increasing temperature of the configurational energy of a system with a dual complex of TAPBs that are spaced by 6, 10 and 16 unit cells from each other ($D = 6, 10, 16$). It should be noted that in all these cases the behaviour of the curves is identical, which leads to the conclusion that there is no mutual influence of TAPB. Such a lack of interaction suggests that at distances between thermal APBs of more than 6 unit cells the TAPBs behave as isolated boundaries.

Thus, comparing the dependences of the configurational energies of a defect-free alloy and an alloy with the dual TAPB complex, spaced apart from each other by 2 unit cells (curves in Fig. 1.16), it is easy to see that the behaviour of the curves is identical. For an alloy with a TAPB complex at a distance of 4 unit cells, the energy values decrease to a temperature of ~400 K, and then coincide with the corresponding values of a defect-free alloy [7, 12, 14]. For the remaining cases, i.e. with further separation of TAPBs from each other, the type of dependence does not change. New features of the temperature dependence do not appear when the distance between the boundaries changes. It can be assumed that the structural changes introduced in this system lead to negligible changes in the configurational energy of the system.

Let us consider the behaviour of integral structural characteristics during thermal cycling: a short-range order parameter (Fig. 1.17 a) and a long-range order parameter (Fig. 1.18 a).

The shape of the temperature dependence of the short-range order parameter [16] on the first coordination sphere of a defect-free system (Fig. 1.17 a) repeats that of the configuration energy dependence (Fig. 1.16 a). The short-range order parameter behaves similarly on the first coordination sphere of a system with a pair of TAPBs (a dual complex) separated by 2, 4, 6, 10, and 16 unit cells in the <100> direction (Fig. 1.17 b and 1.16 b). Therefore, the temperature dependences of the short-range order parameter of both defect-free (Fig. 1.17 a) and defective (Fig. 1.17 b) systems are consistent with the type of change in configurational energies (Fig. 1.16 a and b, respectively).

From the dependences of the long-range order parameter $\eta = \eta(T)$ during thermal cycling (Fig. 1.18 a) in the first cycle, it is easy to see that during the order–disorder phase transition an increase in the temperature to $T = 400$ K does not cause long-range order disturbances. In the temperature range 400–700 K, a gradual decrease in its values is observed, and upon further heating, there is a sharp disordering in the vicinity of $T \approx 800$ K.

With a further increase in temperature, the value of the long-range order parameter tends to zero, which indicates the absence of a long-range order in the alloy. During the disorder–order transition, there is long-range ordering to $T \approx 500$ K. When the alloy is cooled during the disorder–order phase transition, there is no long-range order (Fig. 1.18 a) with a decrease in temperature to $T \approx 500$ K, although short-range ordering begins to appear much earlier (Fig. 1.17 a). A sharp increase in the values of the long-range order parameter is observed when the temperature decreases to about 400 K. With further cooling, its values gradually increase, but do not reach unity. The temperature range of variation of the long-range order parameter (Fig. 1.18 a) is consistent with that of configurational energy (Fig. 1.16 a).

In the second cycle – during heating –, it is easy to see an increase in the values of the long-range order parameter to $T \approx 500$ K, which is caused, as can be assumed, by an increase in diffusion mobility with increasing temperature. A similar increase is observed in subsequent third and fourth cycles, and the maximum values lie

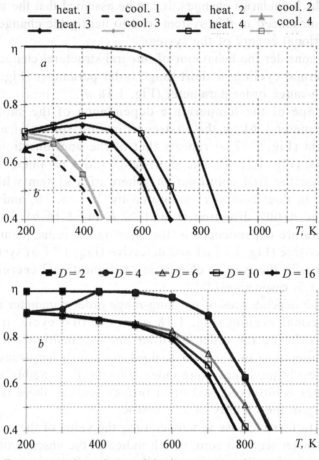

Fig. 1.18. Temperature dependence of the long-range order parameter: *a* – during several successive order–disorder and disorder–order phase transitions during several successive heating-cooling cycles (hysteresis loops in the change in configurational energy during thermal cycling); *b* – during the order–disorder and disorder–order phase transitions in a system with a pair of TAPBs (dual complex) separated by 2, 4, 6, 10, and 16 unit cells in the <100> direction. Hysteresis loops in the variation of the long-range order parameter η during thermal cycling.

in the temperature range 400–500 K. With increasing cycle number, the curve of the temperature dependence of the long-range order parameter, as can be easily seen from Fig. 1.18 *a*, lies lower. A further increase in temperature leads to disordering of the alloy, and hence to a decrease in the values of the long-range order parameter. It should be emphasized that as the cycle number increases, the temperature at which the system is completely disordered decreases.

During the cooling process in the second cycle, it can be seen (Fig. 1.18 *a*) that the non-zero values of the long-range order parameter appear at the same temperatures as in the first cycle. Moreover, the dependence curve $\eta = \eta(T)$ with decreasing temperature lies below the corresponding curve of the first cycle. At $T \to 0$, the values of the long-range order parameter do not reach the corresponding values of the first cycle. The values of the long-range order parameter at $T \to 0$ decrease with the increasing cycle number. With further cycling (increasing *n*), the curves of dependence $\eta = \eta(T)$ with decreasing temperature lie increasingly lower as the cycle number increases.

The behaviour of the long-range and short-range order parameters during thermal cycling indicates that the structural-phase states of the system do not repeat either when the system is heated or when it is cooled. In the process of increasing the temperature of the system, the order–disorder transition, and in the process of lowering the temperature, the disorder–order transition, the system goes through different structural-phase states. As the cycle number increases, these differences decrease, as evidenced by the approximation of the curves of the dependences of the long-range order parameter in the process of thermal cycling.

Let us consider the behaviour of the characteristics of a defective system (modelling CuZn alloy with a pair of TAPB) with increasing temperature during the order–disorder phase transition, i.e. in the process of heating the system. The temperature dependence of the long-range order parameter (Fig. 1.18 *b*) shows that for an alloy with a dual complex of TAPBs separated by 2 unit cells ($D = 2$), at low temperatures the value of the parameter is 1, which corresponds to the ordered alloy with superstructure $B2$. The situation is similar to the case of a defect-free alloy in the first cycle of order–disorder phase transitions (first heating of the system), which can be observed in Fig. 1.18. It can be assumed that at such a distance there is a fairly strong interaction of thermal antiphase boundaries, which leads to healing as a result of diffusion processes of local disordering introduced by the TAPBs.

For an alloy with a dual complex of TAPBs spaced apart from each other by a distance of 4 unit cells ($D = 4$), a very unusual situation is observed. The introduction of a pair of TAPBs naturally leads to a decrease in the long-range ordering, which is manifested by a dip in the curve in the temperature range 200–400 K (Fig. 1.18 *b*). Since the density of defects is quite high, the decrease in the long-range order parameter is very noticeable. At low temperatures ($T < 300$ K), the diffusion mobility of atoms is low, which does not allow the diffusion processes to adapt the structure to the changed conditions for the existence of the system. With increasing temperature ($T > 300$ K), the diffusion mobility of the atoms increases, and the long-range order parameter increases to about 1 at $T = 400$ K. A further increase in temperature, as can be seen from the behaviour of the curve, does not make a significant difference compared with the case of a system with a dual complex of TAPBs separated from each other by a distance of 2 unit cells ($D = 2$).

In other cases ($D = 6, 10, 16$), a monotonic decrease in the ordering is observed with increasing temperature. It should be noted that in all these cases the behaviour of the curves in the temperature range below $T \approx 600$ K is identical, which leads to the conclusion that there is no mutual influence of TAPBs. The absence of interaction suggests that at the distances between thermal APBs of more than 6 unit cells they behave as isolated boundaries.

Let us consider the structure of the CuZn alloy during the order–disorder phase transition in a system with a pair of TAPBs (dual complex) separated by 2 ($D = 2$), 4 ($D = 4$), 6 ($D = 6$), and 10 ($D = 10$) unit cells in the <100> direction with a stepwise increase in temperature. In Fig. 1.19 the dark colour denotes the phases ordered in accordance with the $B2$ superstructure type, and the light colour denotes disordered phases.

It is easy to see that in the case of a model alloy with a dual complex in the form of a pair of TAPBs at a distance of 2 layers, the boundaries annihilate even at low temperatures (200 K). It can be assumed that thermal antiphase boundaries at short distances experience a fairly significant mutual influence. The strong interaction of TAPBs leads to the fact that the structural violation introduced by the introduction of TAPBs is smoothed out by diffusion processes. A further disordering process proceeds with increasing temperature similarly to a defect-free alloy.

The introduction of a pair of TAPBs with an increased distance between thermal APBs in the dual defect up to 3–4 layers leads, as is easily seen from Fig. 1.19, to a decrease in the order in the low-temperature region (200–300 K), which is manifested in the existence of elements of a disordered phase. Since the density of defects is high enough, the decrease in the long-range order parameter is very noticeable. At low temperatures ($T < 300$ K), the diffusion mobility of atoms is low, which does not allow diffusion processes to adapt

T=200 K   T = 300 K   T = 400 K   T = 500 K   T = 600 K   T = 700 K   T = 800 K

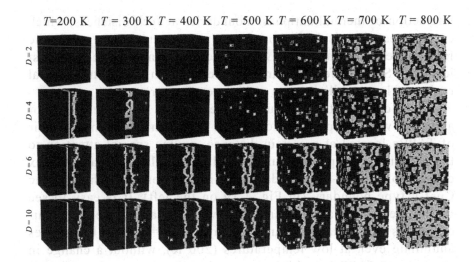

Fig. 1.19. The structural-phase state of the CuZn alloy with a pair of TAPBs in the <100> direction, as a function of temperature and the distance between thermal antiphase boundaries during disordering. Dark colour denotes the phases ordered by the type of $B2$ superstructure, light denotes disordered ones.

the structure to the changed conditions for the existence of the system. This is combined with the behaviour of the long-range order parameter (Fig. 1.18 *b*), which manifests itself in the presence of a dip in the curve in the temperature range 200–400 K. With increasing temperature with increasing diffusion mobility, 'healing' of the order violation occurs (temperature range $T \approx 400$ K), which manifests itself in the form of annihilation of thermal antiphase boundaries (Fig. 1.19). In the curve of dependence of the long-range order parameter this corresponds to its value increasing to 1 (Fig. 1.18 *b*). With increasing temperature ($T > 300$ K), the diffusion mobility of atoms increases, and the long-range order parameter increases to ≈ 1 at $T = 400$ K. A further increase in temperature, as can be seen from the behaviour of the system (Fig. 1.19), does not introduce a significant change compared with the case of a system with a dual complex of TAPBs spaced apart from each other by a distance of 2 unit cells ($D = 2$). Further disordering of the system occurs with increasing temperature similar to a defect-free alloy.

With a further increase in the distance between thermal APBs in the dual defect there is no annihilation of thermal antiphase boundaries (more precisely, 'healing' of the disorder introduced by the TAPBs), even with a significant increase in temperature, and hence diffusion mobility. This is a consequence of the negligible interaction of thermal antiphase boundaries at such distances, which can be treated as isolated boundaries. As a result, a change in the long-range order parameter with increasing temperature is almost identical. Some differences are observed only in the pre-transitional region of low-stability states at high temperatures.

Let us consider the structural-phase features of the behaviour of a system with the presence of a dual defect in the form of a pair of TAPBs located at a certain distance, paying special attention to the vicinity of antiphase boundaries. In Fig. 1.19 the dual complex consists of the Zn–Zn boundary (in Fig. 1.19 on the right) and the Cu–Cu type boundary (in Fig. 1.19 on the left). It is easy to see that in the case of a modelling alloy with a dual complex of a pair of TAPBs at a distance of up to 2 layers ($D = 2$), the boundaries annihilate even at low temperatures (200 K), without a change in the type of TAPBs. At higher temperatures, this trend continues. Apparently, thermal APBs at small distances interact quite strongly, suppressing features as TAPBs. The strong interaction between the TAPBs leads to the fact that the structural distortion introduced by the TAPBs is alleviated by diffusion processes. A further disordering process proceeds with increasing temperature similarly to a defect-free alloy.

The situation changes with increasing distance between thermal APBs. Already in the case of an alloy with a dual complex of TAPBs separated by a distance of 4 unit cells ($D = 4$), a very unusual situation is observed. Already at low temperatures (~200 K) at the Zn–Zn interface (right boundary), significant structural disturbances are visible, which increase with increasing temperature. At the Cu–Cu boundary (left boundary), the structural order disturbances (disordered regions) appear at a higher temperature (~300 K), which can be attributed to a less pronounced structure distortion. At the same temperature, faceting and blurring of the boundaries are observed. A further increase in temperature is accompanied by a disordering process similar to that of a defect-free alloy.

Further separation of the thermal antiphase boundaries from each other ($D = 6$ and 10) leads to the fact that the stage of 'healing' of the structural defect (stage of TAPBs annihilation) during the disordering of the system disappears. Due to a weaker interaction between the thermal APBs, the faceting and blurring stage shifts to higher temperatures (400–500 K). Disordered regions begin to appear throughout the system at 500–600 K, and the shape and size of the boundaries change. A further increase in temperature (~700

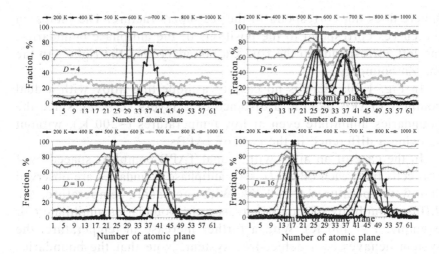

Fig. 1.20. Temperature dependence of the fraction of 'disordered atoms' on the total number in atomic planes parallel to antiphase boundaries during the order–disorder phase transition in a system with a pair of TAPBs (dual complex) separated by 4, 6, 10, and 16 unit cells in the <100> direction.

K) smoothers the shape of the antiphase boundaries. The number of the disordered regions continues increasing (~800 K) throughout the crystal and the antiphase boundaries disappear. At high temperatures almost the entire crystal is disordered, only small-sized domains remaining.

It should be emphasized that the range of change in the boundary structure corresponds to that in the configurational energy of the alloy with TAPBs (Fig. 1.16), i.e. temperature intervals of the existence of low-stability states of the system.

It is known [2–4, 18–23] that, in alloys similar to the one under consideration, during the order–disorder transition, antiphase boundaries are smeared with respect to the positions of atoms in the vicinity of the boundaries [2–4, 18–23] and by the atomic composition of the boundary areas [2–4, 18–23]. Since in this case only diffusion processes are considered without changing the position of the atoms surrounding the boundary, the erosion of thermal antiphase boundaries by atomic composition is studied, paying attention to the size and atomic composition of these boundary regions. We will carry out quantitative estimates, building the dependences of the fraction of 'disordered atoms' on the total number in atomic planes parallel to TAPBs in the <100> direction at different temperatures and distances between antiphase boundaries (Fig. 1.20). By 'disordered atoms' we mean the atoms whose neighbourhood on the first coordination sphere does not correspond to the $B2$ superstructure. The jumps on the curves correspond to the positions of the antiphase boundaries, the left corresponds to the Cu– Cu boundary, and the right – to the Zn–Zn boundary.

In Fig. 1.20, there is no case of a modelling alloy with a dual complex (TAPB pair) at a distance of 2 layers ($D = 2$), since the boundaries annihilate even at low temperatures (~200 K), without showing a difference in the type of TAPBs.

In the case of an alloy with a dual TAPB complex, spaced by 4 unit cells from each other ($D = 4$), antiphase boundaries are observed at low temperatures (~200 K). With increasing temperature, the diffusion mobility of atoms increases, the boundaries annihilate (for example, $T \approx 400$ K). With a further increase in temperature, the system behaves as a defect-free system. Note that the boundaries of the Cu–Cu type (left) and the Zn–Zn type (right) differ both in linear dimensions and in the degree of ordering of the near-boundary region. At the Cu–Cu boundary (left), this region is smaller in linear dimensions and more or less ordered compared to the linear dimensions and ordering of the near-boundary region of the Zn–Zn boundary (right).

The situation changes dramatically with increasing distance ($D = 6, 10, 16$) between TAPBs. At low temperatures (up to ~500 K), the linear dimensions of the near-boundary disordered regions increase with increasing temperature against the background of a general decrease in the system ordering. However, when low-stability states of the system are observed in this temperature range, the linear dimensions of the near-boundary disordered regions stabilize. When the temperature is varied in the temperature range of low-stability states, the sizes of the boundary disordered regions are preserved: for the Cu–Cu type boundary (left), the sizes are of the order of 10 interplanar spacings, and for the Zn–Zn type boundary (right), the sizes are about 12 interplanar distances. Note that the ordering patterns in these regions become similar. It should be noted that the observed features are consistent with the temperature behaviour of internal energy (Fig. 1.16 b), the short- (Fig. 1.17 b) and long-range (Fig. 1.18 b) orders, and the structural-phase state of the system (Fig. 1.19). In all cases, the behaviour of the near-boundary disordered regions in the temperature range of ~500–800 K is similar, which allows us to conclude that there is no mutual influence of TAPBs. The absence of interaction suggests that at distances between thermal APBs of more than 6 unit cells, the APBs behave as isolated boundaries. This behaviour of the system is realized against the background of a monotonous decrease in the general ordering in the system with increasing temperature. Almost the entire crystal transits into a disordered state at the temperatures above 800 K.

Conclusion. It is shown that the presence in the ordered BCC alloy with the $B2$ superstructure of a dual defect in the form of a pair of thermal antiphase boundaries leads to significant structural-phase changes of the system during the order–disorder transition compared to a defect-free system. The presence and nature of the observed features substantially depend both on the temperature and on the distance between the thermal antiphase boundaries.

In an alloy with a dual complex in the form of a pair of thermal antiphase boundaries located at a distance of up to 2 layers (unit cells) in the <100> direction, due to their strong interaction, it can be assumed that the boundaries annihilate even at low temperatures (~200 K) A further order–disorder transformation proceeds similarly to the corresponding process in a defect-free alloy.

In the case of an alloy with a dual TAPBs complex, spaced by up to 3–4 layers, antiphase boundaries are observed at low temperatures (~200 K). With increasing temperature, the diffusion mobility of atoms increases and the boundaries annihilate ($T \approx 400$ K). With a further increase in temperature, the system behaves as a defect-free alloy. Moreover, the Cu–Cu and Zn–Zn type boundaries differ both in linear sizes and in the degree of ordering of the near-boundary regions. Near the Cu–Cu interface, this region is smaller in linear dimensions and less ordered compared to the linear dimensions and ordering of the near-boundary region of the Zn–Zn interface.

The situation changes dramatically with the distance between the TAPBs of more than 6 layers. At low temperatures (up to ~500 K), the linear dimensions of the near-boundary disordered regions increase with increasing temperature against the background of a general decrease in the system's ordering. However, when the low-stability states of the system fall into the temperature range, the linear dimensions of the near-boundary disordered regions stabilize. When the temperature is varied in the temperature range of low-stability states, the sizes of the near-boundary disordered regions are preserved: for the Cu–Cu type boundary, about 10, and for the Zn–Zn type boundary, about 12 interplanar distances. In this case, their ordering in these regions becomes similar.

Blurring and faceting of antiphase boundaries is observed with increasing temperature. Due to the decrease in the interaction of thermal APBs, the faceting and blurring stage shifts to higher temperatures (400–500 K). Disordered regions begin to appear throughout the system at 500–600 K; the shape and size of the boundaries change. A further increase in temperature (~700 K) leads to a change in the shape of the antiphase boundaries to smoother ones. The disordered regions continue to increase (~800 K) throughout the crystal, and the antiphase boundaries disappear. For high temperatures, almost the entire crystal is disordered, only small-sized domains remain. With a further increase in the distance between thermal APBs in the dual defects there is no annihilation of thermal antiphase boundaries (more precisely, 'healing' of the disorder introduced by the TAPBs) does not occur even with a significant increase in temperature, and hence a significant increase in diffusion mobility. This is a consequence of the negligible interaction of thermal antiphase boundaries at such distances, which can be treated as isolated boundaries in this case. As a result, the changes in the long-range order parameter with increasing temperature are almost identical. Some differences are observed only in the pre-transitional region of low-stability states at high temperatures.

In the presence of the TAPBs in the alloy, the first structural order disturbances in the CuZn alloy always appear near the Zn–Zn type boundary.

1.5. Structural-phase low-stability states of BCC alloys with APB complexes during the order–disorder phase transition

Using the Monte Carlo method, we study the effect of complexes of antiphase boundaries in the <100> (a pair of thermal APBs) and <110> (a pair of shear APBs) directions on the features of low-stability structural-phase states and energy characteristics of β-brass (using CuZn alloy as an example) in the order–disorder phase transition.

The formation energy of the antiphase boundary will be found as

$$E^* = \frac{E_{APB} - E}{S},$$

where E is the configurational energy of an ideal defect-free alloy; E_{APB} is the configurational energy of a system with APBs; S is the APB area.

We study the change in low-stability structural-phase states and β-brass configuration energies during the order-disorder phase transition, short-range and long-range order parameters for three model configurations of the CuZn alloy: without any APBs, with a complex of shear APBs in the <110> direction and with a complex of thermal APBs in the <100> direction. The complexes will be built from a pair of antiphase boundaries of the corresponding direction, taking into account the preservation of stoichiometry.

First, we consider a defect-free system (a defect-free CuZn model alloy) during the order–disorder phase transition. In the future, this defect-free system will act as the initial state.

TAPBs normal to the <100> direction are now considered. In this direction, in the $B2$ superstructure, the planes of nodes that are appropriate for Cu and Zn atoms are arranged alternately (see Fig. 1.1). Thermal antiphase boundaries are set by subtracting the planes of Cu or Zn atoms. Subtracting the plane of the Cu atoms, we obtain a thermal antiphase boundary and pairs of the nearest neighbours Zn–Zn (hereinafter, we will call this TAPB a boundary of the Zn–Zn type). Subtracting the plane of the Zn atoms, we obtain the thermal antiphase boundary and the pairs of nearest Cu–Cu neighbours (hereinafter, we will refer to this TAPB as a Cu–Cu type boundary). The Zn–Zn type and Cu–Cu type boundary make up the dual complex in which TAPBs are spaced by a certain distance. Note that with the introduction of such a dual complex, the equiatomic composition of the system does not change. In the presented images of the structural-phase state of the system, the Zn–Zn type boundary will be located on the right, and the Cu–Cu type boundary will be located on the left.

Attention is now paid to the changes in the low-stability structural-phase state of the system as a result of the action of the vacancy diffusion mechanism during the order–disorder phase transition in a system with a pair of TAPBs (dual complex) separated by eight unit cells in the <100> direction.

In the <110> direction the $B2$ superstructure can have only shear APBs. To preserve stoichiometry of the composition, the boundaries are introduced at an angle of 90° in planes passing through the centre of the computational grid. Therefore, in the central part and along the boundaries of the computational block, we have the regions of intersection of the shear boundaries. The intersection of the boundaries is a monoatomic column measuring one unit cell.

When finding the long-range order parameter [17]

$$\eta = \frac{P_A^{(1)} - C_A}{1 - v},$$

it is necessary to calculate the probability. To do this, we determine the number of atoms of sort A, in which the neighbours on the first sphere correspond to the first and second type of domains (parts of the system located on opposite sides of the antiphase boundary). Accordingly, nodes of the first type are considered to be all nodes that are appropriate for atoms of type A, depending on the type of domain:

$$P_A^{(1)} = \frac{N' + N''}{N_1},$$

where N_1 is the number of nodes of the first type; N' is the number of atoms of type A in the nodes of the sublattice of the first type; N'' is the number of atoms of sort A located in the nodes of another sublattice and ordered in accordance with the domain of the second type on the first sphere.

In describing the interatomic interaction, we use the Morse potential parameters given in Table 1.1. The potential values were tabulated as changes in energy depending on interatomic distances.

The study will be carried out in the following sequence: we will consider the structural-phase states and energy characteristics of a defect-free CuZn alloy, an alloy with a complex of antiphase boundaries in the <100> direction (a pair of thermal APBs) and with a complex of antiphase boundaries in the <100> directions (a pair of shear APBs). After a comparative analysis, we reveal the structural-phase and energy features of the low-stability states of the CuZn alloy (β-brass) during the order–disorder phase transition.

Let us consider the CuZn alloy in a defect-free structural state and in states with TAPB and SAPB complexes. The temperature dependence of the average configurational energy during the order–disorder phase transition is shown in Fig. 1.21. It is easy to see that, over the entire temperature range, the energies of states of an alloy with antiphase boundaries are higher than those in the defect-free alloy. Moreover, the values of the average configurational energy of an alloy with a complex of thermal APBs in the <100> direction are significantly higher than the corresponding values of an alloy with shear APBs in the <110> direction. Note that a system with TAPBs in the <100> direction in the temperature range ~300–600

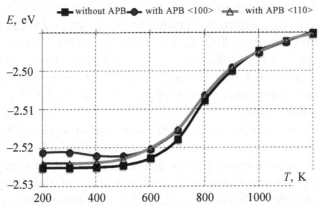

Fig. 1.21. Dependences of the change in the average configurational energy of the CuZn alloy on temperature during the order–disorder phase transition.

K experiences a decrease in configurational energy, which indicates a change in the structure to an energetically more favorable one. A further increase in temperature (above ~600 K) in all cases under study leads to a smooth increase in the value of the configurational energy of the system. At ~1000 K, all configurational variants of the alloy have the same energy values, since the system goes into a completely disordered state. It should be emphasized that in the region of the order–disorder phase transition, the system is in a low-stability state, which is easy to see from the temperature dependences of the energy of various configuration variants of the alloy (Fig. 1.21). Various structural-phase states of the system differ very little in configurational energy, and even minor external influences can easily transfer the system from one state to another.

The average configurational energy of an alloy with a TAPBs complex in the <100> direction does not change at low temperatures (up to 300 K), but there is an insignificant change in the energy difference between the states of the alloy with the TAPBs and SAPBs complexes. In the range from 400 to 500 K, a decrease in the energy difference is observed, which indicates the influence of APBs on the structural and energy characteristics of the alloy during heating. The average configurational energy of the alloy with the SAPB complex at low temperatures also retains its value. However, in the temperature range from 400 to 600 K, an increase in the energy difference is observed. This clearly demonstrates the difference in the effect of various complexes on the average configurational energy of the model alloy. In the vicinity of a temperature of 700 K, a -step- is observed in the temperature dependence of the average configurational energy of the alloy. For higher temperatures, the value of the energy difference gradually decreases and tends to zero. After a temperature of 800 K, i.e. after the order–disorder phase transition, the energy difference is close to zero. This dependence clearly demonstrates that the antiphase boundaries significantly affect the configurational energy of the CuZn alloy up to the phase transition temperature.

Now a pair of TAPBs in the <100> direction or a pair of SAPBs in the <110> direction is introduced into the modelling alloy, and taking into account stoichiometry preservation. The formation energies for each complex in the completely ordered state of the alloy were: Cu–Cu boundaries – 0.0189 eV/$Å^2$, and Zn–Zn boundaries – 0.0365 eV/$Å^2$. Accordingly, in the fully ordered state of the system, the energy of the TAPB complex in the <100> direction is 0.0176 eV/$Å^2$. The energy of the SAPB complex in the direction <110> is, respectively, 0.0036 eV/$Å^2$. The sign is associated with the expression for the introduction of APB energy (E^*), but the presence of an APB naturally leads to an increase in the energy of the alloy.

The temperature dependences of the energy of the TAPB complex in the <100> direction and the energy of the SAPB complex in the <110> direction are shown in Fig. 1.22. The energy of the TAPB complex in the <100> direction is higher than the energy of the SAPB complex in the <110> direction. Of interest is the temperature range 600–900 K, in which, as can be expected from the temperature dependences of the configurational energy of various structural states (Fig. 1.21), structural-phase transformations occur. The same temperature range is of interest because of the behaviour of the energy dependences of the APB complexes (Fig. 1.22), in which a significant change occurs: the energy of the TAPB complex in the <100> direction becomes less than that of the SAPB complex in the <110> direction. This feature is extraordinary, usually the energy of the TAPB complex is higher than that of the SAPB complex. As can be assumed, this inversion is due to the fact that the smearing of SAPBs proceeds through ordering to the B2 superstructure, which, albeit slightly, slows down the general disordering process with increasing temperature. Pay attention to the state of the system at $T = 1000$ K, at which the appearance of complexes of antiphase boundaries leads to a decrease in the energy of the system, which follows from the negative energy values of both SAPBs and TAPBs (Fig. 1.22). This can be explained as follows. At 1000 K, APBs completely disappear; accordingly, the energy difference is already

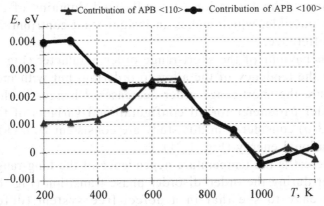

Fig. 1.22. Contribution of APB complexes to the average configuration energy of the alloy with increasing temperature.

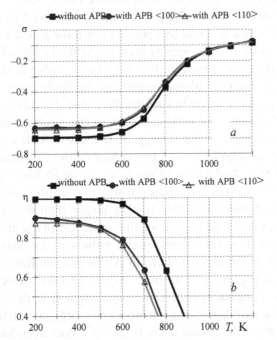

Fig. 1.23. Temperature dependence of parameters of short-range (*a*) and long-range (*b*) orders.

affected not by the initial state of the system, but by random processes caused by the operation of the Metropolis algorithm. It is clear that in the region of higher temperatures the system is in a disordered state; therefore, the dependence curves coincide. In the region of high temperatures, a very small discrepancy between the curves is observed. This may be due to the fact that, at high temperatures, the formation of small-sized domains of different phases is possible, which makes an insignificant contribution to the energy difference.

The behaviour of the short-range order parameter (Fig. 1.23 *a*) indicates the tendency of the alloy to ordering, but with increasing temperature the short-range order decreases. Changes in the short-range order parameter for all configurational variants of the alloy (Fig. 1.23 *a*) considered in this study are consistent with changes in configurational energy (Fig. 1.21).

Of interest is the temperature dependence of the long-range order parameter during the order–disorder phase transition (Fig. 1.23 *b*).

It is easy to see that in a defect-free system (defect-free modelling alloy CuZn) up to $T = 500$ K, there are no long-range order disturbances (η), and in the temperature range ~500–700 K, its value gradually decreases. A sharp decrease is observed in the range $T \approx 700$–900 K, when η tends to zero, which indicates a significant disorder in the system. The appearance of APBs in the alloy leads to a natural result – a decrease in the long-range order in the system during the order–disorder phase transition for both types of APBs (Fig. 1.23 *b*). We draw attention to the fact that the positive values of the energy of the thermal APB complex in the <100> direction are greater than the values of the energy of the shear APB complex in the <110> direction. Deviations of the values of the short-range order parameter from the ideal value (a negative value of the short-range order parameter indicates a tendency towards ordering) for an alloy with a complex of thermal APBs is also larger than the deviations of the short-range order parameter of the shear antiphase boundaries. At the same time, the deviations of the values of the parameter η of a system with a complex of thermal APBs are smaller than those in a system with a complex of shear APBs. In fact, there is some inversion of the dependence. It can be assumed that this is due to the fact that even at low temperatures, the regions of boundary intersection in the alloy with a complex of shear antiphase states transfer to an ordered state, which certainly affects the long-range ordering. The most significant for the long-range order in the system is the very appearance of a structural defect in the form of antiphase atoms; the difference in the plane of their occurrence does not affect the behaviour of η with a change in temperature. Naturally, a system with structural defects is less ordered than a defect-free system, which is manifested in the behaviour of the $\eta = f(T)$ curves: the curve of a defect-free alloy lies higher than the curves of an alloy with the APBs. The presence of a defect in the form of an APB favours the onset of the disordering of the system at lower temperatures: a decrease in the order in the alloy begins already at $T = 300$ K in the case of TAPBs in the <100> direction and at 400 K in the case of SAPBs in the <110> direction. In the temperature range from 400 to 700 K, a stable decrease in the value of the long-range order parameter is observed. With a further increase in temperature, the long-range order tends to zero, which indicates the transition of the alloy to a disordered state. The temperature range of long-range order changes in the alloys is consistent with the range of changes in configurational energy.

Of great interest are changes in the atomic and domain structure of an alloy with an APB complex with increasing temperature of the system. First, we consider the changes in the atomic and domain structure of an alloy with a complex of thermal antiphase atoms in the <100> direction, depending on the temperature during the order–disorder phase transition. Then we will carry out exactly the same investigation of the change in the atomic and domain structure of alloys with a complex of shear antiphase atoms in the <110> direction, depending on the temperature during the order–disorder phase transition. Next, we analyze the influence of the type of APB and the plane of occurrence on the structural-phase features of the alloy depending on temperature versus the order–disorder phase transition temperature.

Let us consider the change in the atomic and domain structure of an alloy with a complex of thermal antiphase atoms in the <100> direction depending on temperature during the order–disorder phase transition, which is shown in Figs. 1.24 and 1.25 respectively. It is easy to see that even at low temperatures (~200 K), significant structural order disturbances occur at the Zn–Zn interface, which increase with increasing temperature. At the Cu–Cu interface, the first disordered regions appear at a higher temperature (~300 K).

In the temperature range 400–500 K there is faceting and blurring of the boundaries. At 500–600 K, disordered regions begin to appear throughout the system; the shape and size of the boundaries change. A further increase in temperature leads to a change in the shape of the antiphase boundaries from flat to three-dimensional (blurred), i.e. the boundary becomes smoother in the sense of a change in ordering in the region of the antiphase boundary ($T \approx 700$ K). The disordered regions continue to increase in their number throughout the system with a further increase in temperature, and the antiphase boundaries

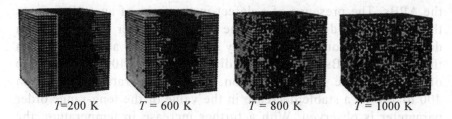

Fig. 1.24. Changes in the atomic structure of an alloy with thermal APB in the process of disordering.

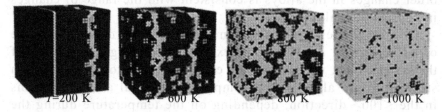

Fig. 1.25. Changes in the domain structure of an alloy with thermal APB during disordering.

disappear ($T \approx 800$ K). At higher temperatures, almost the entire alloy is disordered, only small-sized domains remain. It should be noted that the range of changes in the atomic and domain structure of the system and the low-stability states of the alloy corresponds to the ranges of changes in the configurational energy of the alloy with TAPBs (see Fig. 1.21) and the contribution of APBs to the energy of the alloy (see Fig. 1.22).

Let us consider the change in the atomic and domain structure of an alloy with a complex of shear antiphase atoms in the <110> direction depending on temperature during the order-disorder phase transition, which is shown in Figs. 1.26 and 1.27 respectively.

In this case, structural order disturbances at low temperatures (200–300 K) are observed in the regions of intersection of the antiphase boundaries. With increasing temperature (~400 K), the first disordered regions appear throughout the system but there are no changes in the region of antiphase boundaries. In the range of 500–600 K, faceting and blurring of the boundaries are observed; the number and size of disordered regions increases with increasing temperature. The regions of the boundary intersections tend to transition to an ordered state. A stacking fault defect (a monoatomic column of one of the alloy components measuring one unit cell) is formed in the region of the boundary intersection, which does not

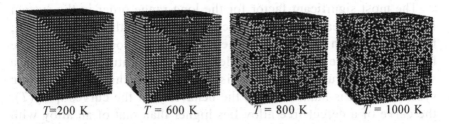

Fig. 1.26. Changes in the atomic structure of an alloy with shear APB during disordering.

Fig. 1.27. Changes in the domain structure of an alloy with shear APB during disordering.

correspond to a simple shear boundary, therefore, its behaviour differs from the behaviour of the SAPBs themselves. These differences provoke an initial rearrangement into an ordered state, while the SAPBs themselves remain stable.

At $T \approx 700$ K, the effect of APBs is still noticeable, but it becomes less pronounced. Naturally, at temperatures above the order–disorder phase transition temperature, the antiphase boundaries disappear, the system is disordered, and only small-sized domains are observed. The situation is similar to that observed in a defect-free alloy and an alloy with a complex of thermal APBs in the <100> direction. Note that the range of variation of the atomic and domain structure of the alloy corresponds to those of the configurational energy of the alloy (see Fig. 1.21) and the contribution of the APB to the energy of the system (see Fig. 1.22).

Conclusion. An analysis of the effect of complexes of antiphase boundaries in the <100> (a pair of thermal APBs) and <110> (a pair of shear APBs) directions on the low-stability structural-phase states of β-brass (using the CuZn alloy as an example) during the order–disorder phase transition showed that the type of antiphase boundaries has a significant effect on the structural-energy state of the entire system.

The most significant factor for the long-range order in the system is the very appearance of a structural defect in the form of APB; the difference in the plane of their occurrence does not affect the behaviour of the long-range order with temperature. Naturally, a system with structural defects is less ordered than a defect-free system, which is manifested in the behaviour of the curves $\eta = f(T)$: the curve of a defect-free alloy lies higher than that of an alloy with APBs. The presence of a defect in the form of an APB promotes the onset of the disordering of the system at lower temperatures: a decrease in the ordering in the alloy begins in the case of TAPBs in the <100> direction at a lower temperature compared with the case of SAPBs in the <110> direction.

In an alloy with a complex of thermal APBs in the <100> direction, the first structural disorder in the CuZn alloy always appears near the Zn–Zn boundary. In an alloy with a complex of shear APBs in the <110> direction, structural order disturbances at low temperatures are observed only in the regions of the boundary intersection. The presence of antiphase boundaries affects the stability of the alloy during its heating. A CuZn alloy without structural defects is more stable than an alloy with APBs. It is shown that the disordering process is accompanied by blurring of the boundaries and their faceting. The contribution of APBs to the disordering process is significant up to the order–disorder phase transition temperature, which is consistent with previous studies [2–15, 18–23].

1.6. Structural-phase features of low-stability pre-transitional states of BCC alloys with complexes of planar defects (antiphase boundaries)

Using the Monte Carlo method, we study the effect of APB complexes in the <100> (a pair of thermal APBs–TAPBs) and <110) (a pair of shear APBs–SAPBs) directions. of APB complexes in the <100> (a pair of thermal APBs–TAPBs) and <110) (a pair of shear APBs–SAPBs) directions on the features of low-stability structural-phase states and energy characteristics of metallic BCC systems (using the example of the traditional CuZn alloy and NiAl intermetallic compound) in the pre-transitional low-stability region of the structural-phase transition during heating.

The features of low-stability pre-transitional structural-phase states and the energy characteristics of β-brass (using the CuZn alloy as an example) are discussed in during the order–disorder structural-phase transformation. A different situation is realized in the case of the NiAl intermetallic compound. It is known that nickel monoaluminide NiAl melts in an ordered state [1]. However, with increasing temperature it undergoes disordering. It is believed that the disordering temperature of the NiAl intermetallic is higher than its melting point. For this reason, we will consider a hypothetical order–disorder transition in order to study the laws of complex atomic ordering aimed at the structural stability and mechanical properties of heat-resistant alloys based on the β-phase of the Ni–Al system, and revealing the influence of disordering processes with increasing temperature on the properties of the intermetallic compound.

We assume that it is the large value of the interatomic interaction forces in the lattice of nickel monoaluminide in comparison with the traditional CuZn alloy which mainly determines the properties of intermetallic compounds. This suggests that from among the mixed covalent, ionic, and metal interatomic bonds in NiAl, only the metal bonding is taken into account, but of a large value compared to that observed in ordinary BCC alloys (for example, CuZn alloy). We study the structural features and energy characteristics of the NiAl intermetallic compound during a hypothetical structural-phase order–disorder transformation depending on the type of thermal and shear antiphase boundaries.

The energy of formation of the antiphase boundary will be found as

$$E^* = \frac{E_{\mathrm{APB}} - E}{S},$$

where E is the configurational energy of an ideal defect-free system; E_{APB} is the configurational energy of the system with APBs; S is the APB area.

We study the change in low-stability structural-phase states and configurational energies of the BCC alloy (CuZn alloy and NiAl intermetallic): without APBs, with a complex of shear APBs in the <110> direction and with a complex of thermal APBs in the <100> direction. The complexes will be built from a pair of antiphase boundaries of the corresponding direction, taking into account the preservation of stoichiometry.

Now consider TAPBs normal to the <100> direction. In this direction, in the $B2$ superstructure the planes of nodes that are appropriate for Cu (Ni) and Zn (Al) atoms are arranged alternately (see Fig. 1.1). Thermal antiphase boundaries are set by subtracting the planes of the Cu (Ni) or Zn (Al) atoms. Subtracting the plane of the Ni atoms, we obtain the thermal antiphase boundary and the pairs of the nearest neighbours Zn (Al)–Zn (Al) (hereinafter, we will refer to this TAPB a boundary of the Zn (Al)–Zn (Al) type). Subtracting the plane of the Zn(Al) atoms, we obtain a thermal antiphase boundary and the pairs of the nearest neighbours Cu (Ni)–Cu (Ni) (hereinafter, we will refer to this TAPB as a boundary of the Cu (Ni)–Cu (Ni) type). The Zn (Al)–Zn (Al) and Cu (Ni)–Cu (Ni) type boundary make up the dual complex in which the thermal antiphase boundaries are spaced apart by a certain distance. Note that with the introduction of such a dual complex, the equiatomic composition of the system does not change. In the presented images of the structural-phase state of the system, the Zn (Al)–Zn (Al) type boundary will be located on the right, and the Cu (Ni)–Cu (Ni) type boundary will be located on the left. Pay attention to the changes in the low-stability structural-phase state of the system as a result of the action of the vacancy diffusion mechanism during a hypothetical order–disorder phase transition in a system with a pair of TAPBs (dual complex) separated by eight unit cells in the <100> direction.

In the <110>direction only shear APBs can exist in the $B2$ superstructure. To preserve stoichiometry of the composition, the boundaries are introduced at an angle of 90° in the planes passing through the centre of the computational grid. Therefore, in the central part and along the boundaries of the computational grid, we have the regions of intersection of the shear boundaries. This is a region of monoatomic composition with a size of 1×1×32 unit cells.

In describing the interatomic interaction, we use the parameters of the Morse potentials given in Table 1.2. The potential values were tabulated as the changes in energy depending on the interatomic distances.

First, let us consider the structural-phase states and energy characteristics of a defect-free system, a system with a complex of antiphase boundaries in the <100> direction (a pair of thermal antiphase boundaries) or with a complex of antiphase boundaries in the <110>direction (a pair of shear APBs) during the order–disorder phase transition. After a comparative analysis, let us determine the structural-phase and energy features of the low-stability states of the CuZn alloy (β-brass) during the order–disorder phase transition and structural and energy features of the low-stability structural-phase states of the NiAl intermetallic during disordering.

Let us consider the CuZn alloy in a defect-free structural state and in the states with TAPB and SAPB complexes. The temperature dependence of the average configurational energy of the CuZn alloy during the order–disorder phase transition is shown in Fig. 1.28 *a*. It is easy to see that, over the entire temperature range, the energies of states of this alloy with antiphase boundaries are higher than that of defect-free ones. Moreover, the values of the average configurational energy of an alloy with a complex of thermal APBs in the <100> direction are significantly higher in the corresponding values of this alloy with shear APBs in the <110> direction. Note that the system with TAPBs in the <100> direction in the temperature range ~300–600 K experiences a decrease in configurational energy, which indicates a change in the structure to an energetically more favourable one. A further increase in temperature (above ~600 K) in all cases considered leads to a smooth increase in the value of the configurational energy of the system. At ~1000 K, all considered configurational variants of the alloy have the same energy values, since the system transits into a completely disordered state. It should be emphasized that in the region of the order–disorder phase transition, the system is in a low-stability state, which is easy to see from the temperature dependences of the energy of various configuration variants of the alloy (Fig. 1.28 *a*). Various structural-phase states of the system differ very little in their configurational energy, and even small external influences can easily transfer the system from one state to another.

While the average configurational energy of the alloy with a TAPB complex in the <100> direction does not change at low temperatures (up to 300 K), but an insignificant change in the energy difference between the states of the alloy with the TAPB and SAPB complexes is observed. In the range from 400 to 500 K, there is a decrease in the energy difference, which indicates the influence of APBs on the structural and energy characteristics of the alloy during heating. The average configurational energy of the alloy with the SAPB complex at low temperatures also retains its value. However, in the temperature range from 400 to 600 K, an increase in the energy difference is observed. This clearly demonstrates the difference in the effect of various complexes on the average configurational energy of the model alloy. In the vicinity of a temperature of 700 K, a 'step' is observed in the graph of the temperature dependence of the average configurational energy of the alloy. For higher temperatures, the value of the energy difference gradually decreases and tends to zero.

At the temperatures above 800 K, i.e. after the order–disorder phase transition, the energy difference is close to zero. This dependence clearly demonstrates that the antiphase boundaries significantly affect the configurational energy of the CuZn alloy up to the phase transition temperature.

Let us consider the NiAl intermetallic compound in a defect-free structural state and in the states with TAPB and SAPB complexes. The temperature dependences of the configurational energy of the NiAl intermetallic during a hypothetical order–disorder phase transition are shown in Fig. 1.28 *b*. It is easy to see that at low temperatures (in the temperature range 200–1800 K) the energies of states of the alloy with antiphase boundaries are higher than those of the defect-free compound. Moreover, the values of the average configurational energy of the alloy with a complex of thermal APBs in the <100> direction. are significantly higher than the corresponding values of the alloy with shear APBs in the <110> direction, which is natural. It should be noted that in the temperature range 1000–2000 K, a significant similarity of the configurational energy of the alloy of all the structural variants under consideration is observed, which can be understood as the realization of the low-stability state of the system. At $T \approx 1200$ K, the energy of the system with a complex of shear APBs becomes comparable with the value of the corresponding characteristic of a system with a complex of thermal APBs. In the temperature range of ~1400–1800 K, the values of the configurational energy of an alloy with shear APBs reveal the energy preference of a system with a complex of shear APBs, i.e. in this temperature range this state is preferable in terms of configurational energy. This can be interpreted as low shear stability of the alloy, i.e. as an opportunity to change its structure to some energetically more preferable one. An

Table 1.2. Morse potential parameters

Type of interaction	α, Å^{-1}	β	D, eV
	CuZn alloy		
Cu–Cu	1.495109	41.598	0.3736
Cu–Zn	1.447832	35.607	0.322
Zn–Zn	1.71223	81.104	0.2189
	NiAl intermetallic		
N–Ni	1.360166	37.72	0.451
Ni–Al	1.073363	17.551	0.6016
Al–Al	1.024939	27.743	0.3724

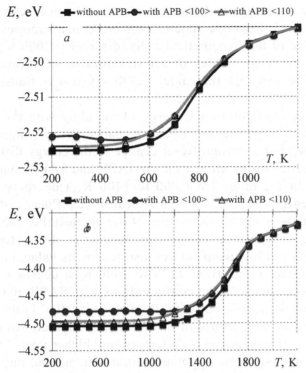

Fig. 1.28. Temperature dependence of the average configurational energy of the CuZn alloy (*a*) and NiAl intermetallic compound (*b*).

increase in temperature above ~1000 K leads to a gradual increase in the configurational energy of both defect-free alloy and the one with a complex of shear antiphase boundaries, but the energy of the former alloy increases more slowly than that of the latter.

Note that a system with TAPBs in the <100> direction in the temperature range ~800–1200 K experiences a decrease in configurational energy, which indicates a possible change in the structure to an energetically more favorable one. A further increase in temperature (above ~1200 K) in all cases considered leads to a smooth increase in the value of the configurational energy of the system. At ~1900 K, all considered configurational variants of the alloy have the same energy values, since the system goes into a completely 'disordered' state. It should be emphasized that in the region of structural-phase transformation (a hypothetical order–disorder transition), the system is in a low-stability state, which is easy to see from the temperature dependences of the energy of various configuration variants of the alloy (Fig. 1.28). Various structural-phase states of the system differ very little in configurational energy, and small external influences can easily transfer the system from one state to another. At temperatures above the temperature of the hypothetical order–disorder (~2000 K) phase transition, the energy values of all structural configurations of monoaluminide NiAl are the same, i.e. the system is transformed into a 'completely disordered' state.

The average configurational energy of the alloy with the TAPB complex in the <100> direction does not change at low temperatures (up to ~800 K), but an insignificant change in the energy difference between the states of the alloy with the TAPB and SAPB complexes is observed. In the range from 800 to 1400 K, a decrease in the energy difference is observed, which indicates the influence of APBs on the structural–energy characteristics of the intermetallic compound during heating. The average configurational energy of a system with a SAPB complex at low temperatures also retains its value, however, in the temperature range from 1200 to 1800 K, a decrease in the energy difference between the configurations with SAPB and TAPB is observed. In the vicinity of 1600 K, there is a 'step' in the graph of the temperature dependence of the average configurational energy of the intermetallic compound. At a temperature higher than ~2000 K, i.e., after an order–disorder hypothetical phase transition, the energy difference is close to zero. This dependence clearly demonstrates that the antiphase boundaries significantly affect the configurational energy of the NiAl intermetallic compound up to the temperature of the structural-phase transformation.

Both general and specific features follow from a comparison of the temperature behaviour of the dependences (Fig. 1.28). Firstly, all pre-transitional low-stability states in the NiAl intermetallic are realized in the region of higher temperatures in comparison with the traditional CuZn alloy. This is due to the stronger interatomic interaction forces in the lattice of nickel monoaluminide in comparison with the traditional CuZn alloy, which mainly determines the properties of intermetallic compounds. The nature of the dependences and their relative positions are similar, but in the case of the NiAl intermetallic compound, they are located in the region of lower system energy values in comparison with the traditional CuZn alloy. The curve of the energy dependence of the CuZn alloy with the complex of thermal antiphase boundaries has a certain dip in the temperature range ~300–600 K (which indicates a change in the structure to an energetically more favorable one), but there is no such dip in the similar dependence of the NiAl intermetallic compound.

The model system (CuZn alloy or NiAl intermetallic) is introduced taking into account stoichiometry of a pair of thermal antiphase boundaries in the <100> direction or a pair of shear antiphase boundaries in the <110> direction. In the CuZn alloy, the formation energies for each complex in the completely ordered state of the alloy were: Cu–Cu boundaries (-0.0189) eV/Å^2, and Zn–Zn boundaries 0.0365 eV/Å^2. Accordingly, in the fully ordered state of the system, the energy of the TAPB complex in the <100> direction is 0.0176 eV/Å^2. The energy of the SAPB complex in the direction <110> is 0.0036 eV/Å^2. In the NiAl intermetallic compound in a fully ordered state, the energy of formation of a complex of thermal APBs in the <100> direction was 0.1037 eV/Å^2, and that of the complex of shear APBs in the <110> direction – 0.0233 eV/Å^2. Naturally, the energy of the complex of thermal APBs is about 4 times higher than the energy of the complex of shear APBs.

Figure 1.29 shows the dependences of the contributions of the APB complexes to the average configuration energy of the (a) CuZn alloy and (b) NiAl intermetallic compound with increasing temperature. The temperature dependences of the energy of the TAPB complex in the <100> direction and the energy of the SAPB complex in the <110> direction of the CuZn alloy are shown in Fig. 1.29 a. The energy of the TAPB complex in the <100> direction is higher than the energy of the SAPB complex in the <110> direction, which is natural. Of interest is the temperature range of 600–900 K, in which, as can be expected from the temperature dependences of the configurational energy of various structural states (Fig. 1.28 a), structural-phase transformations occur. The same temperature range is of interest because of the behaviour of the energy dependences of the APB complexes (Fig. 1.29 a), in which a significant change occurs: the energy of the TAPB complex in the <100> direction becomes less than the energy of the SAPB complex in the <110> direction. This feature is unconventional, usually the energy of the TAPB complex is greater than the energy of the SAPB complex. It can be assumed that this inversion is due to the fact that the smearing of SAPB proceeds through ordering to the $B2$ superstructure, which, though slightly, slows down the general disordering process with increasing temperature. Let us pay attention to the state of the system at $T = 1000$ K, at which the appearance of complexes of antiphase boundaries leads to a decrease in the energy of the system, which follows from the negative energy values of both SAPBs and TAPBs (Fig. 1.29 a). This can be explained as follows. At 1000 K,

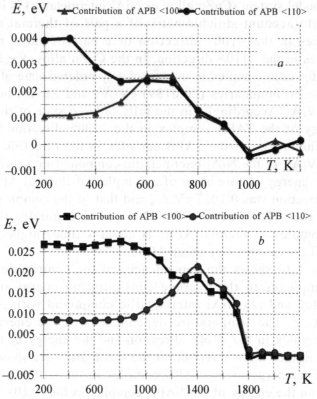

Fig. 1.29. Contribution of APB complexes to the average configurational energy of (*a*) CuZn alloy and (*b*) NiAl intermetallic with increasing temperature.

APBs completely disappear; accordingly, the energy difference is already affected not by the initial state of the system, but by random processes caused by the operation of the Metropolis algorithm. It is clear that in the region of higher temperatures the system is in a disordered state; therefore, the dependence curves coincide. In the region of high temperatures, a very small discrepancy of the curves is observed. This may be due to the fact that, at high temperatures, the formation of small-sized domains of different phases is possible, which makes an insignificant contribution to the energy difference.

A slightly different situation is in the case of NiAl intermetallic compound (Fig. 1.29 *b*). It is easy to see that at temperatures up to about 1200 K the energy of shear antiphase boundaries is much lower than the thermal energy. We note that the contribution of the SAPB complex to the average configurational energy of the system practically does not change to a temperature of ~800 K, and the contribution of the TAPB complex to $T \approx 1000$ K changes insignificantly. In the range 900–1300 K, a significant decrease in the energy difference is observed, which indicates the influence of the presence of antiphase boundaries rather than their type on the structural-energy characteristics upon heating. In the vicinity of the temperatures of 1300 and 1500 K, the temperature dependence of the contribution of the TAPB complex shows local features – wells. This can be explained by a change in the nature of the disordering process from disordering near the boundaries to disordering throughout the system. It is interesting that in the temperature range from 1400 to 1700 K for a system with a complex of shear APBs, the contribution of defects to the average configurational energy of the system becomes greater than that for a system with a complex of thermal APBs. At higher temperatures, the curves tend to zero, which indicates the absence of the influence of the initial configuration on the state of the system. The contribution of SAPBs is maximum at $T \approx 1400$ K; at higher temperatures, it decreases. The appearance of the peak in the dependence at $T = 1400$ K can be explained by the fact that the system with the SAPB complex begins to disorder at lower temperatures than the defect-free alloy. A further increase in temperature reduces this difference. Comparing the changes in the average configurational energy with the temperature dependence (Fig. 1.28 b), this can be understood as the low shear stability of the alloy, i.e. as an opportunity to change its structure to some more energetically more preferable.

Thus, complexes of antiphase boundaries significantly affect the structural-phase states and configurational energies of the NiAl intermetallic compound, while the energy characteristics of the alloy depend not only on temperature, but also on the type of antiphase boundaries and their plane of occurrence.

It is easy to see the general features of the temperature behaviour of the contributions of the APB complexes to the average configuration energy of the system (Fig. 1.29); however, in the case of the NiAl intermetallic compound, this occurs at higher temperatures in comparison with the case of the traditional CuZn alloy, which is natural. In both cases, the course of the curves is similar: at low temperatures, the energy of the complex of thermal antiphase boundaries significantly exceeds the energy of shear APBs. An inversion is observed in a certain temperature range of low-stability pre-transitional states: the contribution of the shear antiphase complex exceeds the contribution of the thermal antiphase complex. At higher temperatures in the region of low-stability pre-transitional states, the system becomes indifferent to the type of antiphase boundaries: the contribution of the shear antiphase complex is equal to the contribution of the thermal antiphase complex. In the temperature range above the temperatures of the structural-phase transformation, virtual states are observed in which the system with both shear and thermal APB has certain energy preferences compared to a defect-free system.

The behaviour of the short-range order parameter on the first coordination sphere of both the CuZn alloy (Fig. 1.30 a) and the NiAl intermetallic compound (Fig. 1.30 b) indicates the tendency of the alloy or intermetallic compound to ordering (a negative value of the short-range order parameter indicates a tendency towards ordering). However, with increasing temperature, the short-range order decreases. A system with a complex of shear APBs is more prone to ordering than a system with a complex of thermal APBs. Naturally, the structural-phase transition in the NiAl intermetallic compound occurs at higher temperatures than the order–disorder phase transition in the CuZn alloy. Changes in the short-range order parameter for all considered configurational variants of the alloy (Fig. 1.30) are consistent with changes in configurational energy (Fig. 1.28).

It is of interest to compare the temperature dependences of the long-range order parameter of the traditional CuZn alloy (Fig. 1.31 a) during the order–disorder phase transition and intermetallic monoaluminide NiAl (Fig. 1.31 b) upon heating, especially in the region of low-stability pre-transitional states of the system. First, we consider the CuZn alloy, then the NiAl intermetallic compound, and then compare them.

It is easy to see (Fig. 1.31 a) that in a defect-free system (defect-free CuZn model alloy) up to $T = 500$ K, long-range order (η) is not disturbed, while in the temperature range ~500–700 K, its value gradually decreases. A sharp decrease is observed in the range $T \approx 700$–900 K as the η value tends to zero, which indicates a significant disorder in the system. The appearance of an APB in the alloy leads to a natural result – a decrease in the long-range order in the system during the order–disorder phase transition for both types of APB (Fig. 1.31 a).

We draw attention to the fact that the positive values of the energy of the thermal APB complex in the <100> direction are greater than the values of the energy of the shear APB complex in the <110> direction. Deviations of the values of the short-range order parameter from the ideal value (a negative value of the short-range order parameter indicates a tendency towards ordering) for an alloy with a complex of thermal APBs is also larger than the deviations of the short-range order parameter of the shear APBs (Fig. 1.30 *a*). At the same time, the deviations of the values of the parameter η of the system with a complex of thermal APBs are smaller than the deviations of the values of η of the system with a complex of shear APBs (Fig. 1.31 *a*). In fact, there is some inversion of dependence. It can be assumed that this is due to the fact that even at low temperatures, the regions of boundary intersection in the CuZn alloy with a complex of shear antiphase atoms transfer to an ordered state, which, of course, affects the long-range order. The most significant for the long-range order in the system is the appearance of a structural defect in the form of antiphase boundaries,

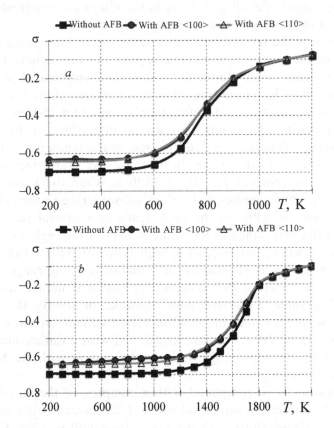

Fig. 1.30. Temperature dependence of the short-range order parameter of the CuZn alloy (*a*) during the order-disorder phase transition and the NiAl intermetallic compound (*b*) upon heating.

the difference in the plane of their occurrence does not affect the behaviour of η with temperature as much. Naturally, a system with structural defects is less ordered than a defect-free system, which is manifested in the behaviour of the dependence curves η = $f(T)$: the curve of a defect-free alloy lies higher than the curves of an alloy with an APB. The presence of a defect in the form of an APB promotes the onset of disordering of the system at lower temperatures: a decrease in the order in the alloy begins already at T = 300 K in the case of a TAPB in the <100> direction and at 400 K in the case of an SAPB in the <110> direction. In the temperature range from 400 to 700 K, a stable decrease in the value of the long-range order parameter is observed. With a further increase in temperature, the long-range order tends to zero, which indicates the transition of the alloy to a disordered state. The temperature range of the long-range order (Fig. 1.31 a) in the alloys is consistent with the range of the configurational energy (Fig. 1.28 a).

Of interest is the temperature dependence of the long-range order parameter of the NiAl intermetallic compound in the region of low-stability pre-transitional states, for which the hypothetical process of the order–disorder phase transition is considered (Fig. 1.31 b).

It is easy to see that in a defect-free system (defect-free model NiAl intermetallic compound) there are no distortions of the long-range ordering (η) up to T = 1100 K, and in the temperature range ~1100–1600 K its value gradually decreases. A sharp decrease is observed in the interval $T \approx$ 1600–1800 K when the value of η tends to zero, which implies a significant disordering in the system. The appearance of APBs in the alloy leads to a natural result – a decrease in the long-range order in the system in the region of low-stability pre-transitional states for both types of APBs (Fig. 1.31 b). Note that at low temperatures the value of the long-range order parameter in a system with a complex of TAPBs is higher than the corresponding value in a system with a complex of SAPBs. However, at $T \approx$ 700 K, the η values in both considered variants of the system with complexes of defects become equal, and with a further increase in temperature, the long-range order in the system with the SAPB complex becomes higher.

In an intermetallic compound with a complex of thermal APBs in the <100> direction, structural order disturbances begin at low temperatures (from 300 K), in the range from 900 to 1200 K, the preservation of the long-range order parameter is observed, a further increase in temperature leads to a sharp decrease in its value. The

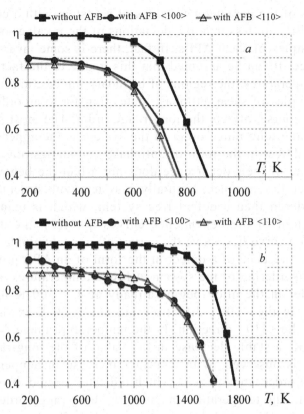

Fig. 1.31. Temperature dependence of the long-range order parameter of the CuZn alloy (*a*) during the order–disorder pre-transitional process and the NiAl intermetallic compound (*b*) during heating.

configuration of a monoaluminide with a complex of shear APBs in the <110> direction, on the contrary, demonstrates a stable value of the long-range order parameter up to 1000 K, and when the temperature rises to about 1400 K, the value of the long-range order parameter gradually decreases. In a system with APB complexes, the long-range order parameter naturally decreases faster than in a defect-free alloy. Note that in the NiAl intermetallic compound, the positive values of the energy of the complex of thermal APBs in the <100> direction are larger than those of the complex of shear APBs in the <110> direction. Deviations of the values of the short-range order parameter from the ideal value (a negative value of the short-range order parameter indicates a tendency towards ordering) in an alloy with a complex of thermal APBs are also larger than those of the short-range order parameter in the case of a system with a complex of shear APBs. At the same time, at low temperatures, the deviations of the parameter η values of a system with a complex of thermal APBs are less than the deviations of the η values of a system with a complex of shear APBs. In fact, there is some inversion of the dependence. It can be assumed that this is due to the fact that even at low temperatures, the regions of boundary intersection in the alloy with a complex of shear antiphase states transfer to an ordered state, which, of course, affects the long-range order. The most significant factor for the long-range order in the system is the appearance of a structural defect in the form of antiphase boundaries, the difference in the plane of their occurrence does not affect the behaviour of η with temperature as much. Naturally, a system with structural defects is less ordered than a defect-free system, which is manifested in the behaviour of the dependence curves $\eta = f(T)$: the defect-free intermetallic curve lies above the curves of monoaluminide NiAl with APBs. The presence of a defect in the form of an APB promotes the onset of disordering of the system at lower temperatures: a decrease in the order in the alloy begins already at $T \approx 300$ K in the case of TAPBs in the <100> direction and at ~1100 K in the case of SAPBs in the <110> direction. In the temperature range from ~1100 to ~1800 K, a stable decrease in the value of the long-range order parameter is observed. With a further increase in temperature, the long-range order tends to zero, which indicates a rapid disordering in the system. The temperature range of the long-range order changes in the system (Fig. 1.31 b) is consistent with the range of changes in configurational energy (Fig. 1.28 b).

Comparing the behaviour of the long-range order parameter of defect-free states of the traditional CuZn alloy (Fig. 1.31 a) and NiAl intermetallic compound (Fig. 1.31 b), we can conclude that the dependences $\eta = f(T)$ in both cases are similar: in the low-temperature region, $\eta \approx 1$, i.e. the average atomic long-range order is maintained in the system. Naturally, in the intermetallic compound this remains until the range of higher temperatures. The disordering occurs in the pre-transitional region of low-stability states (in both cases, the fall of the long-range order parameter η from ≈1 to ≈0.4 occurs in the range of ~300 K before the structural-phase transformation).

However, the situation changes significantly when structural complexes appear in the system. For an alloy and an intermetallic compound with shear APBs, in both cases the decrease in η values occurs monotonously with increasing temperature. In the case of a system with thermal APBs, the types of dependences of decreasing η values of the alloy (Fig. 1.31, a) and intermetallic compound

(Fig. 1.31 *b*) differ. Firstly, if the dependence $\eta = f(T)$ in the alloy is monotonic with increasing temperature, then in the case of an intermetallic it exhibits a nonmonotonous behaviour in the pre-transitional low-stability region (in the vicinity of $T \approx 1000$ K). This may indicate the features of atomic ordering in this system in these states. Moreover, since the 'system-averaged' characteristic is used, this behaviour $\eta = f(T)$ suggests the possibility of local structural-phase transformations in this state region.

Let us pay attention to the mutual position of the $\eta = f(T)$ curves of the system with thermal and shear APBs both in the case of the traditional CuZn alloy (Fig. 1.31 *a*) and NiAl intermetallic compound (Fig. 1.31 *b*). It is easy to see significant differences in the region of pre-transitional low-stability states.

In the case of the traditional CuZn alloy (Fig. 1.31 *a*), the $\eta = f(T)$ curves of the system with thermal and shear APBs are monotonically decreasing, and the curve corresponding to the alloy with TAPBs always lies above the curve corresponding to the alloy with SAPBs. The curves are tangent to each other in a certain temperature range 400–500 K, revealing the system's indifference in this temperature range with respect to atomic ordering in a system with planar structural defects in the form of antiphase boundaries. The differences in the values of the long-range order parameter in both cases are small at all the temperatures considered, i.e., long-range order in the system is not critical to the type of APBs. Abn important fact is the presence of the structural defects in the form of antiphase boundaries in the system as such.

In the case of the NiAl intermetallic compound (Fig. 1.31 *b*), the $\eta = f(T)$ curves of a system with thermal and shear APBs behave differently. The curve corresponding to the system with SAPBs decreases monotonically, while the curve corresponding to the system with TAPBs has a dip in the vicinity of $T \approx 1000$ K. The curves intersect twice (at $T \approx 700$ and 1300 K). The differences in η values are significant and by far exceed the differences in η values of the CuZn alloy. At low temperatures (up to $T \approx 700$ K), the long-range order in the system with TAPBs exceeds the long-range order in the system with SAPBs, but in the temperature range 700–1300 K the situation is reversed: the long-range order in the system with SAPBs exceeds the long-range order in the system with TAPBs. The inversion is noticeable and significant: the sensitivity of the system in a given temperature range with respect to atomic ordering in a system with planar structural defects in the form of antiphase boundaries is revealed. The system appreciably responds not only to the appearance of structural defects in the form of antiphase boundaries, but it is also critical to their type. Hence, it can be assumed that the NiAl intermetallic compound in the region of pre-transitional low-stability states is critical in the long-range order with respect to the structural state of the system. It is not a long-range order convergence in different structural states of the system, but a change in the atomic distribution, i.e. system structure. It can be assumed that in the case of NiAl intermetallic in the region of pre-transitional low-stability states, a change in the values of the long-range order parameter is a consequence of structural changes in the system. This distinguishes the NiAl intermetallic compound from the traditional CuZn alloy, in which, as can be assumed, atomic disordering (an order-disorder phase transition) causes structural-phase changes in the system.

A comparative analysis of the features of disordering processes in the traditional CuZn alloy and NiAl intermetallic with increasing temperature in the region of low-stability pre-transitional states leads to a very interesting result. The order–disorder phase transition in the CuZn alloy occurs as a result of disordering in the system, while in the NiAl intermetallic compound, the long-range order changes as a result of the structural-phase transition.

It is appealing to look at the changes in the atomic and domain structure of the system (CuZn alloy or NiAl intermetallic compound) with a complex of APBs with increasing temperature. First, we consider the changes in the atomic and domain structure of a system with a complex of thermal APBs in the <100> direction, versus the temperature in the region of low-stability pre-transitional states with increasing temperature. Then we will carry out exactly the same consideration of the changes in the atomic and domain structure of a system with a complex of shear APBs in the <110> direction as a function of temperature. Next, we analyze the influence of the type of APBs and the plane of their occurrence of antiphase on the structural-phase features of the system depending on temperature in the region of low-stability pre-transitional states.

We study the changes in the atomic and domain structure of the system (CuZn alloy or NiAl intermetallic compound) with a complex of thermal antiphase atoms in the <100> direction depending on temperature (Figs. 1.32 and 1.33, respectively), paying particular attention to pre-transitional low-stability states. Note that in all the presented images of the structural-phase state of the system,

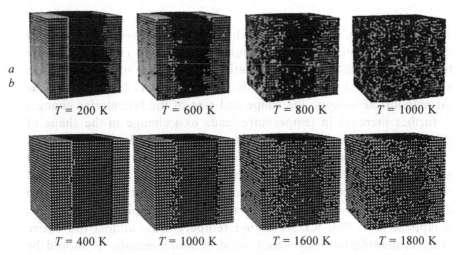

Fig. 1.32. Changes in the atomic structure of (*a*) CuZn alloy and (*b*) NiAl intermetallic complex of thermal antiphase atoms in the <100> direction in the region of low-stability pre-transitional states.

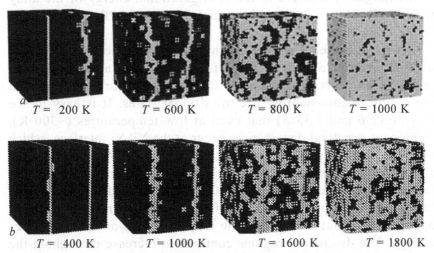

Fig. 1.33. Changes in the domain structure of (*a*) CuZn alloy and (*b*) NiAl intermetallic with a complex of thermal antiphase atoms in the <100> direction in the region of low-stability pre-transitional states.

the Zn (Al)–Zn (Al) type boundary is located on the right, and the Cu (Ni)–Cu (Ni) type boundary is on the left.

Let us consider the change in the atomic and domain structure of the CuZn alloy with a complex of thermal antiphase boundaries in the <100> direction, depending on the temperature during the order–disorder phase transition. It is easy to see (Fig. 1.32 *a* and 1.33 *a*) that even at low temperatures (~200 K), significant structural disturbances occur at the Zn–Zn interface, which increase with increasing temperature. At the Cu–Cu interface, the first disordered regions appear at a higher temperature (~300 K). In the temperature range 400–500 K, faceting and blurring of the boundaries are observed. At 500–600 K, disordered regions begin to appear throughout the system; the shape and size of the boundaries change. A further increase in temperature leads to a change in the shape of the antiphase boundaries from flat to bulk (blurred),

In other words, the boundary becomes smoother in the sense of a change of the ordering in its vicinity ($T \approx 700$ K). The disordered regions continue to increase in number throughout the system with a further increase in temperature, and the antiphase boundaries disappear ($T \approx 800$ K). At higher temperatures, almost the entire alloy is disordered, only small-sized domains remain. It should be noted that the range of changes in the atomic and domain structure of the system and the low-stability states of the alloy corresponds to the ranges of changes in the configurational energy of the alloy with TAPBs (Fig. 1.28 b) and the contribution of APBs to the energy of the alloy (Fig. 1.29 b).

Let us consider a change in the atomic and domain structure of NiAl intermetallic with a complex of thermal antiphase atoms in the <100> direction, relative to the temperature variations in the region of low-stability pre-transitional states. It is easy to see (Fig. 1.32 b and 1.33 b) that even at low temperatures (~300 K), significant structural disturbances occur at the Ni–Ni interface, which increase with increasing temperature. At the Al–Al interface, the first disordered regions appear at a higher temperature (~600 K). In the temperature range 600–1400 K, faceting and blurring of the boundaries are observed. At ~1000 K, disordered regions begin to appear throughout the system; the shape and size of the boundaries change. The disordered regions continue to increase throughout the system with a further increase in temperature, and the antiphase boundaries disappear ($T \approx 1600$ K). At higher temperatures, almost the entire alloy is disordered, only small-sized domains remain. We note that the range of changes in the atomic and domain structure of the system and low-stability states corresponds to the range of changes in the configurational energy of the alloy with TAPBs (Fig. 1.28 b) and the contribution of APBs to the energy (Fig. 1.29 b).

Let us consider the change in the atomic and domain structure of a system (CuZn alloy or NiAl intermetallic compound) with a complex of shear APB in the <110> direction depending on temperature (Fig.

1.34 and 1.35, respectively), paying particular attention to pre-transitional low-stability states.

We study the changes in the atomic and domain structure of the CuZn alloy with a complex of shear APBs in the <110> direction as a function of temperature during the order–disorder phase transition (Figs. 1.34 *a* and 1.35 *a*, respectively). In this case, structural order disturbances at low temperatures (200–300 K) are observed in the regions of intersection of antiphase boundaries.

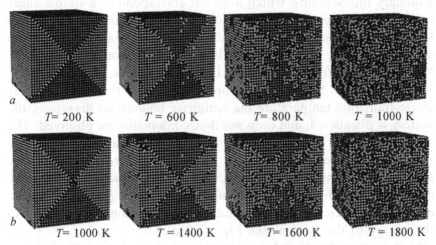

Fig. 1.34. Changes in the atomic structure of (*a*) CuZn alloy and (*b*) NiAl intermetallic with a complex of shear APBs in the <110> direction in the region of low-stability transition states.

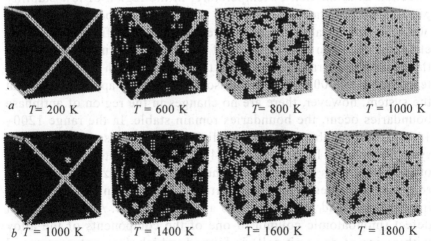

Fig. 1.35. Changes in the domain structure of (*a*) CuZn alloy and (*b*) NiAl intermetallic with a complex of shear APB in the <110> direction in the region of low-stability transition states.

With increasing temperature (~400 K), the first disordered regions appear throughout the system; but no changes occur in the region of antiphase boundaries. In the range of 500–600 K, faceting and blurring of the boundaries are observed; the number and size of disordered regions increases with increasing temperature. The regions of the boundary intersections tend to transition to an ordered state. A stacking fault defect (a monoatomic column of one of the alloy components the size of one unit cell) is formed in the region of the boundary intersections, which does not correspond to a simple shear boundary, and therefore its behaviour differs from that of the SAPBs themselves. These differences induce an initial rearrangement into an ordered state, while the SAPBs themselves remain stable. At $T \approx 700$ K, the influence of APBs is still noticeable, but it becomes less pronounced. Naturally, at temperatures above the order–disorder phase-transition temperature the antiphase boundaries disappear, the system is disordered, and only small-sized domains are observed. The picture is similar to that observed in a defect-free alloy and an alloy with a complex of thermal APBs in the <100> direction. Note that the range of changes in the atomic and domain structure of the alloy corresponds to the ranges of changes in the configurational energy of the alloy (see Fig. 1.28 a) and the contribution of the APBs to the energy of the system (see Fig. 1.29 a).

Let us consider the change in the atomic and domain structure of NiAl intermetallic with a complex of shear APBs in the <110> direction as a function of temperature, in the region of low-stability pre-transitional states (Fig. 1.34 b and 1.35 b, respectively). Up to 600 K no structural disturbances are observed. The first insignificant changes occur in the regions of the boundary intersections in the temperature range from 600 to 1000 K. With an increase in temperature (~1100 K), the first disordered regions appear throughout the system, however, there are no changes in the region of antiphase boundaries occur, the boundaries remain stable. In the range 1200–1400 K, the number and size of disordered regions throughout the system increases. In the range of 1400–1500 K, faceting and blurring of the boundaries are observed; the number and size of disordered regions increases with temperature. The regions of boundary intersections tend to transition to an ordered state. A stacking fault defect (monoatomic column of one of the components of the alloy with a size of one unit cell) is formed, which does not correspond to a simple shear boundary, therefore, its behaviour differs from that of the SAPBs themselves. These differences provoke an initial rearrangement into an ordered state, while the SAPBs themselves remain stable. At $T \approx 1600$ K, the effect of APBs is still noticeable, but it becomes less pronounced. Naturally, at temperatures above the temperature of a hypothetical order–disorder phase transition, the antiphase boundaries disappear, the system is disordered, and only small-sized domains are observed. The situation is similar to that observed in a defect-free alloy and an alloy with a complex of the thermal APBs in the <100> direction. We note that the range of changes in the atomic and domain structure of the intermetallic compound corresponds to that in the configurational energy of the alloy (see Fig. 1.28 b) and the contribution of APBs to the energy of the system (see Fig. 1.29 b).

Conclusion. An analysis of the influence of complexes of antiphase boundaries in the <100> (a pair of thermal APBs) and <110> (a pair of shear APBs) directions on low-stability structural-phase states of a BCC alloy (using CuZn alloy and an intermetallic monoaluminide NiAl as an example) showed that the presence of antiphase boundaries has a significant impact on the structural-phase state and energy characteristics of the entire system.

The most significant factor for the long-range order in the system is the appearance of a defect in the form of APBs; the difference in the type and plane of their occurrence does not significantly affect the behaviour of the long-range order with temperature. Naturally, a system with structural defects is less ordered than a defect-free system, which is manifested in the behaviour of the curves $\eta = f(T)$: the curve of a defect-free system lies higher than the curves of a system with APBs. The presence of a defect in the form of an APB contributes to the onset of disordering of the system at lower temperatures: a decrease in order begins in the case of thermal APBs in the <100> direction at a lower temperature compared with the case of shear APBs in the <110> direction.

From a comparative analysis of the features of disordering in the traditional CuZn alloy and NiAl intermetallic with increasing temperature in the region of low-stability pre-transitional states, it follows that while the order–disorder phase transition in the CuZn alloy occurs as a result of disordering in the system, then the long-range order changes in the NiAl intermetallic as a result of structural-phase transformation.

In a BCC system (a traditional CuZn alloy or NiAl intermetallic compound) with a complex of thermal APBs in the <100> direction, the first structural order disturbances always appear near the antiphase boundary separating the neighbouring Zn–Zn atoms in the CuZn alloy and the neighbouring Ni–Ni atoms in the NiAl intermetallide. In a system with a complex of shear APBs in the <110> direction, structural order disturbances at low temperatures are observed only in the regions of boundary intersection. The presence of antiphase boundaries affects the stability of the system during heating: a system without structural defects is more stable than a system with APBs. It is shown that the disordering process is accompanied by blurring of the boundaries and their faceting. The contribution of APBs to the disordering process is significant up to the temperature of the structural-phase transition, which is consistent with the previous studies [2–15, 18–23].

1.7. Interaction of a complex of thermal antiphase boundaries during an order–disorder phase transition in a CuZn alloy

Let us discuss the change in pre-transitional low-stability states, configurational energy, short-range and long-range order parameters in the CuZn model alloys with a dual complex of SAPBs and with a pair of TAPBs for all configurations of the model alloy. The distance between TAPBs will change from 0 (dual complex) to 16 layers of unit cells, the step of changing the distance between TAPBs will be 2 layers.

For calculations, we use the Monte Carlo Metropolis algorithm. The interactions between different pairs of atoms of the alloy components will be specified using the semi-empirical Morse pair potential. The Morse potential parameters describing the interatomic bonds of the pairs of components A–A, B–B, and A–B are taken from Table 1.1.

The modelling system consists of $32 \times 32 \times 32$ unit cells (65536 atoms). At the boundaries of the system, we will set periodic boundary conditions, which effectively corresponds to an infinite system.

Figure 1.36 shows the temperature dependences of configurational energy (a) and short-range order parameter (b) on the first coordination sphere for various distances between TAPBs (D) in the direction <100> during the order–disorder phase transition.

It can be seen (Fig. 1.36 a) that the curve of the system without defects, i.e., TAPBs are found naturally below the corresponding curves of the alloy containing defects. The values of configurational

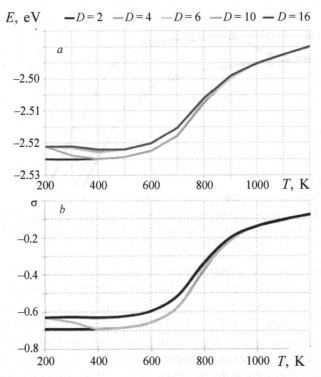

Fig. 1.36. Temperature change in configurational energy (*a*) and short-range order on the first sphere (*b*) as a function of the distance between TAPBs in the <100> direction upon heating during the order–disorder phase transition.

energy for an alloy with a TAPB complex spaced by distance of 2 ($D = 2$) unit cells are similar to the energy values for a defect-free alloy. For an alloy with a TAPB complex at a distance of 4 unit cells, the energy decreases to a temperature of 400 K, then increases and coincides with that in a defect-free alloy, starting at a temperature of $T \approx 800$ K. Further, an increase in the distance between the TAPBs does not change the temperature dependence of the configurational energy.

The changes in the short-range order parameter σ (Fig. 1.36 *b*) indicate that the strongest tendency of the alloy towards ordering is observed at low temperatures (the most negative values of σ are observed at low temperatures). As the temperature rises, the tendency to ordering decreases (negative values decrease in their absolute value). The temperature dependences σ of various configurations of the system are consistent with the those of the configurational energy of these configurations.

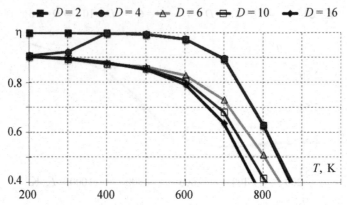

Fig. 1.37. Temperature dependence of the long-range order parameter in the order-disorder phase transition process in the system with a pair of TAPBs (dual complex) separated by 2, 4, 6, 10 and 16 unit cells in the direction <100>.

The temperature dependence of the long-range order η (Fig. 1.37) shows that for $D = 2$ at low temperatures the value of the parameter η is 1, which corresponds to an ordered alloy with the $B2$ superstructure. For $D = 4$, the long-range order parameter at low temperatures is less than unity (some disordering is introduced by the TAPBs), however, η increases to 1 at $T = 400$ K and then completely coincides with the case $D = 2$. With increasing distance between TAPBs in other experiments, the value of the long-range order gradually decreases with increasing temperature, the rate of decrease depends on the distance between the boundaries. The greater the distance, the faster the violation of long-range order.

Let us consider the behaviour of the characteristics of a defective system (model CuZn alloy with a pair of TAPBs) with increasing temperature during the order–disorder phase transition during heating. The temperature dependence of the long-range order parameter (Fig. 1.37) shows that for an alloy with a dual complex of TAPBs separated by 2 unit cells ($D = 2$), at low temperatures the parameter value is 1, which corresponds to an ordered alloy with the $B2$ superstructure. The situation is similar to the case of a defect-free alloy in the first cycle of order–disorder phase transitions. It can be assumed that a sufficiently strong interaction of thermal antiphase boundaries is present at this distance, which leads to 'healing' as a result of diffusion processes of local disordering introduced by the TAPBs.

For an alloy with a dual complex of TAPBs spaced apart from each other by a distance of 4 unit cells ($D = 4$), a very unusual situation is observed. The introduction of a pair of TAPBs naturally leads to a decrease in the long-range order, which is manifested by a dip in the curve in the temperature range 200–400 K (Fig. 1.37). Since the density of defects is quite high, the decrease in the long-range order parameter is very noticeable. At low temperatures ($T <$ 300 K), the diffusion mobility of atoms is low, which does not allow diffusion processes to adapt the structure to the changed conditions for the existence of the system. With increasing temperature ($T >$ 300 K), the diffusion mobility of atoms increases, and the long-range order parameter also increases to about 1 at $T = 400$ K. A further increase in temperature, as can be seen from the behaviour of the curve, does not introduce a significant difference compared with the case of a system with a dual complex of TAPBs separated from each other by a distance of 2 unit cells ($D = 2$).

In other cases under consideration ($D = 6, 10, 16$), a monotonic decrease in ordering is observed with increasing temperature. It should be noted that in all these cases the behaviour of the curves in the temperature range below $T \approx 600$ K is identical, which leads to the conclusion that there is no mutual influence of TAPBs. Such a lack of interaction suggests that at distances between thermal APBs of more than 6 unit cells, behave as isolated boundaries.

Let us consider a change in the domain structure of the CuZn alloy during the order–disorder phase transition (Fig. 1.38). The ordered phases are marked with dark and light colour, respectively.

It is easy to see that in the model alloys with a dual complex of APBs and with a pair of TAPBs at a distance of up to 2 layers, the boundaries annihilate at low temperatures (~200 K). The further process proceeds similarly to a defect-free alloy. An increase in the distance to 3–4 layers also leads to annihilation of the boundaries, but only in the temperature range of ~300–400 K. At distances between the boundaries of more than 3–4 layers, even at low temperatures (~200 K), significant distortions of strucural order are visible at the Zn–Zn interface. They increase with temperature; for 300 K, the first disordered regions appear at the Cu–Cu interface. In the range 400–500 K, faceting and blurring of the boundaries are observed. Disordered regions begin to appear throughout the computational grid at ~500–600 K; the shape and size of the boundaries change. The range of changes in the boundary structure corresponds to that in the configurational energy of the alloy with TAPBs (Fig. 1.36 *a*). A change in the shape of the antiphase boundaries is observed to become smoother (~700 K), which indicates the contribution of temperature to the general disordering process. The disordered

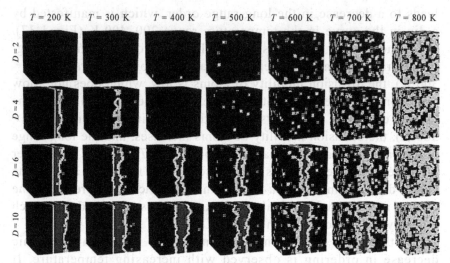

Fig. 1.38. Changes in the domain structure of the CuZn alloy with a TAPB pair in the <100> direction depending on the distance between the boundaries during disordering. The dark colour indicates the ordered phases; the light colour indicates the disordered ones.

regions continue to increase throughout the crystal at ~800 K, the antiphase boundaries disappear. For high temperatures, almost the entire crystal is disordered, only small-sized domains remain.

With an increase in temperature, blurring of the boundaries is observed [2–15, 18–23]. Blurring was evaluated by plotting the number of 'disordered atoms' in the planes parallel to TAPBs (Fig. 1.39). The term 'disordered atom' refers to the atoms whose neighbourhood on the first coordination sphere does not correspond to the $B2$ superstructure. The jumps on the curves correspond to the positions of the antiphase boundaries.

Figure 1.39, where $D = 4$, shows that for temperatures above ~200 K, the boundaries annihilate. A typical curve is characterized by two peaks, the left peak corresponds to the Cu–Cu type boundary, and the right one corresponds to Zn–Zn boundary. An increase in temperature to ~400 K leads to blurring of the boundaries. Distortions of the structural order throughout the crystal are noticeable at ~500 K. The boundaries continue to blur. A higher temperature (~600 K) leads to an increase in the number of disordered atoms throughout the computational block. With a further increase in temperature to ~700 K, the boundaries become more blurred, disordering is observed in all layers. Almost the entire crystal goes into a disordered state at temperatures above ~800 K.

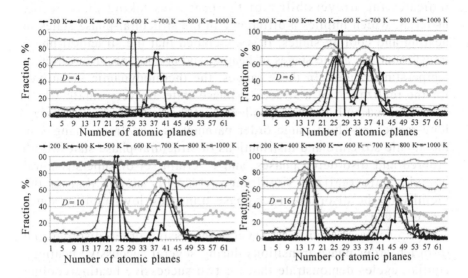

Fig. 1.39. Temperature dependence of the number of 'disordered atoms' in layers parallel to antiphase boundaries.

Conclusion. It has been shown that in the model alloys with a dual APB complex and with a pair of TAPBs at a distance of up to 2 layers, the boundaries annihilate at low temperatures (~200 K). The further process with increasing temperature proceeds similarly to that in a defect-free alloy. An increase in the distance to 3–4 layers also leads to annihilation of boundaries in the temperature range of ~300–400 K. Blurring and faceting of antiphase boundaries is observed with increasing temperature with a further increase in the distance. In the presence of an APB alloy in the CuZn alloy, the first structural order disturbances always appear near the Zn–Zn boundary.

Summary

Using the Monte Carlo Metropolis algorithm, the features of thermal cycling of pre-transitional low-stability structural-phase states of BCC alloys have been considered (using traditional CuZn alloy and NiAl intermetallic compound) in various situations: during one or several thermal cycles (during several successive heating-cooling cycles)), in the presence of complexes of planar defects (shear and thermal antiphase boundaries), and in the course of the interaction of complexes of the thermal antiphase boundaries.

It has been shown that in all cases, as a result of each heating and cooling cycle, a hysteresis is observed, the presence of which indicates the irreversibility of the processes taking place in the material, which implies a difference in the structural-phase states in heating and cooling stages. It is concluded that the structural-phase transformations at the stages of heating and cooling occur in different temperature ranges. In these ranges, the thermodynamic incentives for the realization of a structural-phase state are very small, which can be traced both in the dependences of the configurational energy, long-range and short-range order parameters, and from changes in the atomic structure and distributions of structural-phase states. Both ordered and disordered phases, and a certain set of superstructural domains, are realized simultaneously. This means that in the vicinity of the disorder–order phase transition low-stability states are observed.

The results of studies of several successive order–disorder and disorder–order phase transitions during several successive heating–cooling cycles demonstrate that for two successive heating-cooling cycles at the same temperature, the structural-phase states differ both at the heating stage and at cooling stage. These differences are not dramatic, but occur on each cycle, and decrease as the cycle number increases. Actually, the system undergoes training with the tendency towards forming a certain steady-state sequence of structural-phase states.

It is shown that the influence of planar defects (antiphase boundaries) on the disordering process is significant up to the temperature of the structural-phase transformation. The most significant for the long-range order is the appearance of a defect itself; the difference in the type of APBs and their plane of occurrence does not affect the behaviour of the long-range order with temperature. The type of antiphase boundaries significantly affects the structural and energy characteristics of the system at the temperatures below the phase transformation temperature. Naturally, a system with structural defects is less ordered than a defect-free system. The presence of a defect contributes to the onset of disordering of the system at lower temperatures: a decrease in the order in the alloy begins in the case of thermal APBs at a lower temperature compared with the case of shear APBs. In an alloy with a complex of thermal APBs in the <100> direction, the first structural disorder in the CuZn alloy always appears near the Zn–Zn boundary. In an alloy with a complex of shear APBs in the <110> direction, structural order disturbances at low temperatures are observed only in the regions of boundary intersection.

The presence of antiphase boundaries affects the alloy stability during its heating. A CuZn alloy without structural defects is more stable than the one with APBs. The disordering process is accompanied by blurring of the boundaries and their faceting.

The presence of a dual defect in the form of a pair of thermal antiphase boundaries in an ordered BCC alloy with a $B2$ superstructure (using the CuZn alloy as an example) leads to significant structural-phase features of the system during the order-disorder transition compared to a defect-free system. The presence and nature of the observed features substantially depend both on temperature and on the distance between thermal antiphase boundaries. In this case, the Cu–Cu and Zn–Zn type boundaries differ both in linear sizes and in the degree of ordering of the near-boundary regions. At the Cu–Cu interface, this region is smaller in linear dimensions and less ordered than near the Zn–Zn interface. At low temperatures, the linear dimensions of the boundary disordered regions increase with increasing temperature against the background of a general decrease in the system's ordering degree.

In the region of low-stability states of the system, the sizes of the near-boundary disordered regions are preserved: for the Cu–Cu boundary, about 10 interplanar distances, and for the Zn–Zn boundary, about 12. In this case, the ordering in these regions becomes similar, blurring and faceting of antiphase boundaries are observed, moreover, the first disordered regions always appear near the Zn–Zn-type boundary.

Using the traditional CuZn alloy and the NiAl intermetallic as examples, the influence of the APB separated complexes (pairs of shear APBs and pairs of thermal APBs) on low-stability pre-transitional states of BCC alloys is considered. It has been shown that, in the region of low-stability structural-phase states, the energy of formation of a complex of thermal APBs is higher than that necessary for a complex of shear APBs. The contribution of APBs to the disordering process is significant up to the temperature of the structural-phase transformation. The most significant factor for the long-range order is the appearance of the defect itself; the difference in the type of antiphase boundaries and their plane of occurrence does not affect the behaviour of the long-range order with temperature. The type of antiphase boundaries significantly affects the structural and energy characteristics of the system at temperatures below the phase transformation temperature. Naturally, a system with structural defects is less ordered than a defect-free system. The presence of a defect contributes to the onset of disordering of the system at lower temperatures: a decrease in the order in the alloy begins in the case of TAPBs at a lower temperature compared with the case of shear APBs. In the BCC system with the TAPB complex, the first order disturbances always appear near the antiphase boundary separating larger atoms (Zn–Zn in the CuZn or Al–Al alloy in the NiAl intermetallic compound). In an alloy with a complex of shear APBs, order disturbances at low temperatures are observed only in the regions of boundary intersection. The presence of antiphase boundaries affects the stability of the alloy during heating. It has been shown that the disordering process is accompanied by blurring of the boundaries and their faceting. A comparative analysis of the features of disordering processes in BCC systems (traditional CuZn alloy and NiAl intermetallic) has demonstrated that an increase in temperature in the region of low-stability pre-transitional states implies that an order–disorder phase transition in the Cu–Zn alloy occurs as a result of disordering in the system, while in the NiAl intermetallic compound the long-range order changes as a result of the structural-phase transformation. To sum up, disordering predominates in the traditional CuZn alloy and the structural-phase transformation does so in the NiAl intermetallic compound.

References

1. Kositsyn S.V., Kositsyna I.I. // Advances in metal physics. - 2008. - V. 9. - P. 195–258.
2. Potekaev A.I., Starostenkov M.D., Kulagina V.V. The effect of point and planar defects on structural-phase transformations in the pre-transitional low-stability region of metal systems / Ed. A.I. Potekaev. - Tomsk: NTL Publishing House, 2014 .-- 488 p.
3. Koneva N.A., Trishkina L.I., Potekaev A.I., Kozlov E.V. Structural-phase transformations in weakly stable states of metallic systems during thermosilic interaction / under the general. ed. A.I. Potekaev. - Tomsk: NTL Publishing House, 2015 .-- 344 p.
4. Potekaev A.I., Naumov I.I., Kulagina V.V., et al. Low-stability metallic-based nano-structures / ex. ed. A.I. Potekaev. - Tomsk: Scientific Technology Publishing House, 2018 .-- 236 p.
5. Chaplygin P. A., Potekaev A. I., Chaplygin A. A. et al. // Izv. Univ. Fizika. - 2015. - V. 58. - No. 4. - P. 52–57.
6. Chaplygina A.A., Chaplygin P.A., Starostenkov M.D. et al. // Fundamental problems of modern materials science. - 2016. - V. 13. - No. 3. - P. 403–407.
7. Potekaev A.I., Chaplygina A.A., Chaplygin P.A. et al. // Izv. Univ. Fizika. - 2016. - V. 59. - No. 5. - P. 3–8.
8. Potekaev A.I., Chaplygina A.A., Starostenkov M.D. et al. // Izv. Univ. Fizika. - 2012. - V. 55. - No. 7. - P. 78–87.

9.   Potekaev A.I., Chaplygina A.A., Starostenkov M.D. et al. // Izv. Univ..Fizika. - 2012. - V. 55. - No. 11. - P. 7–16.
10.   Potekaev A.I., Chaplygina A.A., Starostenkov M.D. et al. // Izv. Univ. Fizika. - 2013. - V. 56. - No. 6. - P. 14–22.
11.   Potekaev A.I., Klopotov A.A., Kozlov E.V., Kulagina V.V. // Izv. Univ. Fizika. - 2011. - V. 54. - No. 9. - P. 59–69.
12.   Potekaev A.I., Chaplygina A.A., Kulagina V.V. et al. // Izv. Univ.. Fizika. - 2017. - V. 60. - No. 2. - P. 16–26.
13.   Potekaev A.I., Klopotov A.A., Trishkina L.I. et al. // Bulletin of the Russian Academy of Sciences. Ser. physical - 2016. - V. 80. - No. 11. - P. 1576-1578.
14.   Chaplygina A.A., Potekaev A.I., Chaplygin P.A. et al. // Fundamental problems of modern materials science. - 2016. - V. 13. - No. 2. - P. 155−161.
15.   Chaplygina A.A., Potekaev A.I., Chaplygin P.A. et al. // Izv. Univ.. Fizika. - 2016. - V. 59. - No. 10. - P. 13–22.
16.   Iveronova V.I., Katsnelson A.A. Short-range order in solid solutions. - M .: Nauka, 1977 .-- 253 p.
17.   Krivoglaz M.A., Smirnov A.A. Theories of ordered alloys. - M .: Fizmatgiz, 1958.- 388 p.
18.   Potekaev A.I., Kulagina V.V. // Izv. Univ.. Fizika. - 2011. - V. 54. - No. 8. - P. 5–22.
19.   Potekaev A.I., Kulagina V.V. // Izv. Univ.. Fizika. - 2009. - V. 52. - No. 8/2. - P. 456–459.
20.   20. Potekaev A.I. // Fiz. Met. Metalloved. - 1986. - V. 61. - No. 2. - P. 254–264.
21.   Potekaev A.I. // Phys. Stat. Sol. (a). - 1992. - V. 134. - P. 317–334.
22.   Potekaev A.I., Klopotov A.A., Kozlov E.V., Kulagina V.V. // Izv. Univ.. Fizika. - 2011. - V. 54. - No. 9. - P. 59–69.
23.   Potekaev A.I., Kulagina V.V., Klopotov A.A. // Izv. Univ.. Fizika. - 2011. - V. 54. - No. 4. - P. 11–18.

Influence of various factors on low-stability pre-transitional structural-phase states of NiAl intermetallic compound

Currently, special attention is drawn to materials with a complex of special properties and the ability to preserve them in extreme conditions. Ordered alloys and intermetallic compounds show a great promise for practical application due to a range of unique physical and physico-mechanical properties, such as strength, heat resistance, shape memory, superelasticity, magnetic properties, etc. The complex of unique properties is mainly determined by structural-phase state of the metal system.

A characteristic representative of such materials is the NiAl intermetallic compound of the Ni–Al system. Nickel aluminide is being actively studied as a promising material for the aerospace industries. Useful characteristics of NiAl are its high melting point, relatively low density, good chemical resistance, high thermal conductivity, high strength, and metal-like properties. A characteristic feature of the Ni–Al alloys system is a high ordering energy. The NiAl intermetallic compound and its substitutional solid solutions have a high degree of long-range order, which is maintained in the entire temperature-concentration region of their existence up to the melting point.

The NiAl intermetallic monoaluminide has a BCC lattice ordered by the $B2$ type (CsCl) with the $Pm3m$ space group, in which two simple cubic sublattices of nickel and aluminum can be distinguished. The Ni–Al system is characterized by a large variety of the atomic sizes and electronic structure types. The electronic structure of NiAl (β-phase) is characterized by strong hybridization of d(Ni)–p(Al) bonds along the <111> direction between the nearest-neighbour atoms in Ni–Al vapours (strong covalent component), depletion by electrons of the Ni and Al positions in the <100> direction between the neighbours of the second coordination sphere, and an increased electron density between the nearest-neighbour Ni–Al atoms in the <111> direction (weak ionic bond). These directions of bonding also prevail over the metal bondd [1]. The presence of mixed covalent, ionic, and metal interatomic bonds in NiAl determines a large unit cell volume and a long Burgers vector, a decrease in the independent equivalent slip systems, and complex reactions of dislocation interactions with each other, and various boundaries and stacking faults, determines slip localization, and complicates the transfer of deformation across the boundary. The properties of β-alloys are largely determined by the strength of interatomic interaction forces in the lattice of nickel monoaluminide [1].

Creating materials with a set of target properties and a given structural-phase state of the system is far from being a trivial task. The study of the properties and structural-phase states of materials by experimental methods is laborious and costly. Moreover, it is often very difficult to identify the mechanisms of the ongoing physical and chemical processes. In this situation, computer modelling is quite useful, since many processes and phenomena are difficult to observe in a real experiment. For this reason, systematic studies of the structural-phase states of metal systems by computer simulation methods attract particular attention and are of particular value, since in this case it is possible to reveal the physicochemical processes and phenomena occurring in the system [2, 3].

Classical studies of low-stability pre-transitional states have been performed on β-alloys of the Ni–Al system [4–16]. An important direction is to investigate the competition and the mutual influence of parallel processes (ordering and decay of a β-solid solution; ordering and microdisintegration; ordering and martensitic transformation) and regulation of complex atomic ordering processes in order to increase the structural stability and mechanical properties of heat-resistant intermetallic compounds on the basis of the β-phase of the Ni–Al system. It is known that the nickel monoaluminide is characterized by a high melting point (1638°C) and a large value of the heat of formation. NiAl crystals exhibit strong elastic anisotropy and related anisotropy of properties in contrast to the structures with a disordered BCC lattice [1].

Naturally, the properties of alloys are associated with their structural phase state, ultimately, the properties and structure of phases, structural defects. The study of the properties and structural-phase state of materials by computer simulation methods allows us to study in detail the mechanisms of the ongoing physico-chemical processes [2–18]. A characteristic representative of the structural defects of intermetallic compounds is the complex of antiphase boundaries (APBs), which are a special type of plane defects. A characteristic feature of the shear APBs is that all atoms located on one side of the boundary plane are shifted by a vector connecting the atoms of two sublattices relative to atoms on the other side of the boundary. For the $B2$ superstructure such a shift corresponds to a change in the sorts of all atoms by the sorts of the opposite component. In alloys with a $B2$ superstructure, shear APBs (SAPBs) are formed in planes with an even sum of Miller indices, and thermal (TAPBs) with an odd one. TAPBs in the $B2$ superstructure are formed predominantly in the planes of the cube and the octahedron. An important characteristic of an AFB is the energy of its formation. The lower the energy (surface tension) of the APB formation the larger the distance to which the dislocations propagate..

2.1. Applied approximations and the model used

The structural-phase features of low-stability pre-transitional states and the energy characteristics of intermetallic BCC compounds are investigated here by the example of NiAl intermetallide during heating and cooling in various situations: during thermal cycling in the course of structural-phase transformations, during stepwise cooling, under the influence of vacancy concentration, variations of the alloy composition and grain size, and the use of different APB complexes (pairs of shear APBs in the <110> direction and pairs of thermal APBs in the <100> direction).

It is known that NiAl melts in an ordered state [1]. Disorder, however, occurs with increasing temperature. It is believed that the disordering temperature of the NiAl intermetallic is higher than its melting temperature. For this reason, hypothetical order–disorder transitions during heating and disorder–order transitions during cooling are examined in order to study the laws of complex atomic ordering –disordering with a purpose to increase the structural stability and mechanical properties of heat-resistant alloys based on the β-phase of the Ni–Al system, to identify the effect of disordering processes with increasing temperature on the properties of this intermetallic compound. We assume that it is precisely the large value of the interatomic interaction forces in the NiAl lattice which determines mainly the properties of β-alloys. This suggests that, due to the presence of a mixed covalent, ionic, and metal interatomic bonds in NiAl, only the metal bond (which is quite large compared to that observed in ordinary BCC alloys) is taken into account.

Here we will pay special attention to low-stability pre-transitional states. The temperature range of such states will be estimated from the analysis of hypothetical order–disorder transitions during heating and disorder–order transitions during cooling in the NiAl intermetallic.

To this aim, we consider the structural-phase states and energy characteristics of a defect-free NiAl intermetallic compound at $T = 0$ K. A defect-free system will subsequently act as the initial state. The lattice parameters of the starting configuration of the alloy are determined by the minimum configurational energy, which is searched by the gradient descent method. Next, we study the change in low-stability structural phase states and configurational energies of the NiAl intermetallic compound, depending on the situation in the system under consideration. To activate the diffusion process, for example, one vacancy is introduced randomly into the system, which corresponds to a concentration of $\sim1.5 \cdot 10^{-5}$. Only the vacancy diffusion mechanism is considered. The dynamic or kinetic component is present only in jumps of atoms to vacant sites.

When studying the states of this system with complexes of antiphase boundaries, we will firstly: consider thermal antiphase boundaries (TAPB) normal to the <100> direction.

In this direction, in the $B2$ superstructure, the planes of nodes that are appropriate for Ni and Al atoms alternate (see Fig. 1.1). Thermal antiphase boundaries are set by subtracting the planes of the Ni or Al atoms. Subtracting the plane of the Ni atoms, we obtain a thermal antiphase boundary and pairs of the nearest Al–Al neighbours (hereinafter, we will call such a TAPB as an Al–Al type boundary). Subtracting the plane of the Al atoms, we obtain the thermal antiphase boundary and pairs of the nearest Ni–Ni neighbours (hereinafter, we will refer to this TAPB as a Ni–Ni type boundary). The Al–A and the Ni–Ni type boundary make up a dual complex in which the thermal antiphase boundaries are spaced apart by a certain distance. Note that with the introduction of such a dual complex, the equiatomic composition of the system does not change. In the presented images of the structural phase state of the system, the Al–Al boundary is located on the left, and the Ni–Ni boundary on the right. Attention should be given to changes in the low-stability structural phase state of the system as a result of the action of the vacancy diffusion mechanism during the hypothetical order–disorder phase transition in a system with a pair of TAPBs (dual complex) separated by eight unit cells in the <100> direction.

In the <110> direction the $B2$ superstructure contains only shear APBs. To preserve stoichiometry of the the composition, the boundary is made at an angle of 90° in planes passing through the centre of the computational grid. Therefore, in the central part and along the boundaries of the computational grid, we have the regions of intersection of the shear boundaries. The boundary intersection area is a region of monoatomic composition with a size of 1 × 1 × 32 unit cells.

For calculations, we use the Monte Carlo Metropolis algorithm. To activate the diffusion process, one vacancy is randomly introduced into the system, which corresponds to a concentration of ~$1.81\cdot10^{-5}$. We use only the vacancy diffusion mechanism. The dynamic or kinetic component is present only in jumps of atoms to vacant sites.

An ordered BCC structure with a $B2$ superstructure is considered (see Fig. 1.1). The model (computational grid unit) includes 32 × 32 × 32 unit cells (65536 atoms), and we use periodic boundary conditions, which effectively corresponds to an infinite system with a long period. We assume that a central pair interaction is realized between the atoms. In this case, we will define the interatomic interaction in the form of a central pair Morse potential, which has proved itself quite well in such problems [4, 5, 9–18]:

$$\varphi(r_{ij}) = D_{KL}\,\beta_{KL}\,e^{-\alpha_{KL}r_{ij}}\left(\beta_{KL}\,e^{-\alpha_{KL}r_{ij}} - 2\right),$$

where α_{KL}, β_{KL}, D_{KL} are the parameters of the potentials describing the bonds of atoms of varieties K–L; r_{ij} is the distance between atoms. The configurational energy of the system is calculated as

$$E = 1/2\sum_{i=1}^{N}\sum_{j=1}^{M}\varphi\left(r_i - r_j\right),$$

where $r_i - r_j$ are the radius vectors of atoms i and j; N is the number of atoms in the system; M is the number of nearest neighbours, which includes the atoms of the three coordination spheres of interaction. The applied values of the Morse potential parameters are given in Table 2.1.

For calculations, the Monte Carlo Metropolis algorithm is used. Details of the modelling technique are given in Sec. 1.1 of the previous chapter.

For each temperature, 5×10^{6} iterations were performed, the temperature change step was 100 K.

In the study, special attention is paid to the changes in configurational energy, short-range and long-range order parameters, structure-phase low-stability states during heating (hypothetical disordering) and cooling (hypothetical ordering).

The short-range order parameter on the i-th sphere will be determined in the Cowley approximation [19]:

$$\sigma_i^{AB} = 1 - \frac{P_i^{AB}}{C_B}$$

where C_B where is the concentration of atoms of component B; P_i^{AB} is the probability of the formation of the A–B bond for the i-th atom in the i-th coordination sphere.

The long-range order parameter (averaged over the system) will be calculated in the Gorsky–Bragg–Williams approximation [20]:

$$\eta = \frac{P_A^{(1)} - C_A}{1 - v},$$

where $P_A^{(1)}$ is the probability of filling of nodes of the first type with the atoms of component A; C_A is the concentration of atoms of component A in the alloy; v is the concentration of nodes of the first type.

2.2. Structural-phase transformations in BCC intermetallic compounds during thermal cycling

The structural-phase characteristics of the BCC alloy during thermal cycling were investigated by a Monte Carlo simulation. When describing the interatomic interaction, the Morse potential parameters given in Table 2.1 were used. The potential values were tabulated as changes in energy depending on interatomic distances.

In the process of thermal cycling, the average configurational energy per atom (Fig. 2.1), the long-range order parameter (Fig. 2.2 *a*) and the short-range order parameter (Fig. 2.2 *b*) were studied.

From Fig. 2.1 it is easy to see that at temperatures below 900 K the average configurational energy per atom does not change either during heating or cooling, but during cooling its value is noticeably higher. A gradual increase in the energy value in the temperature range from 1000 to 1400 K with its growth is replaced by a sharp increase in the range from 1400 to 1900 K, which corresponds to the disordering processes (Fig. 2.2), i.e., in this temperature range, an order–disorder phase transition occurs. It is easy to see that the temperature range of variation of the short-range and long-range order parameters (Fig. 2.2) is consistent with the temperature range of variation of configurational

Table 2.1. Morse potential parameters for NiAl alloy

Type of interaction	α	β	D
Ni–Ni	1.360166	37.72	0.451
Ni–Al	1.073363	17.551	0.6016
Al–Al	1.024939	27.743	0.3724

energy (Fig. 2.1). As a result of the heating and cooling cycle, a kind of hysteresis is observed (Fig. 2.1), the presence of which during thermal cycling indicates the irreversibility of the processes and suggests the difference in the structural phase states at the stages of heating and cooling. This is reflected in the behaviour of the system-average characteristics with temperature, namely, the long-range order (Fig. 2.2 *a*) and short-range (Fig. 2.2 *b*) order parameters. It is easy to see that the numerical values of these characteristics differ significantly at the same temperature in the heating and cooling stages. Thus, even the behaviour of the average characteristics of the system indicates the irreversibility of the processes occurring during the order–disorder phase transition during thermal cycling.

During heating, as can be seen from Fig. 2.2, in the course of the order–disorder phase transition, there are no long-range order disturbances to a temperature of 1200 K, and then to a temperature of 1600 K the long-range order value gradually decreases. A sharp jump in the curve of the dependence of the long-range order parameter on temperature would occur at $T = 1700$ K, which indicates a rapid disordering in the system. It can be assumed that, upon heating, the order–disorder phase transition occurs in a certain temperature range near $T = 1700$ K. With a further increase in temperature, the long-range order tends to zero, which indicates disordering in the system.

During cooling, there is no long-range order to $T = 1600$ K, while, the short-range order appears even at temperatures below 1800 K (Fig. 2.2 *b*). A sharp increase in the value of the long-range order parameter is observed in the temperature range 1100–1400 K, which suggests that an order–disorder phase transition occurs in this temperature range. Comparing the obtained data with the temperature interval of the heating stage, we can conclude that the system had been supercooled in the cooling stage. At temperatures below 800 K, the values of the long-range order parameter in the alloy do not change; the system is in an ordered state.

Since, as a result of the heating and cooling cycle, a peculiar hysteresis is observed (Fig. 2.1), the presence of which during thermal cycling indicates the irreversibility of the processes, the structural-phase

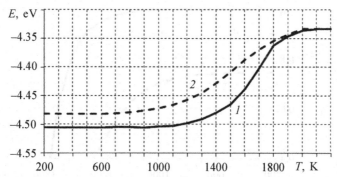

Fig. 2.1. Average configurational energy per atom during sequential heating (curve 1) and cooling (curve 2).

states at the heating and cooling stages would be different. To prove this, we analyze the atomic and phase structures of the system during heating and cooling. Figures 2.3 and 2.4 show the atomic structure of the alloy as a function of temperature during the order–disorder and disorder–order phase transitions, respectively.

When heated (Fig. 2.3) to a temperature of 1000 K, the alloy is ordered, with an increase to $T = 1200$ K the first disordered regions appear. With a further increase in temperature to 1600 K, the number and size of regions with disturbances of the superstructural arrangement of atoms increase. At temperatures above 1800 K, the superstructural arrangement of the atoms is substantially distorted; upon reaching a temperature of 1900 K, the alloy is completely disordered.

When the system is cooled (Fig. 2.4) with a decrease in temperature to 1800 K, regions appear that are ordered in accordance with the $B2$ superstructure. At $T = 1600$ K, the number and size of ordered regions increases, and at temperatures below 1300 K, the alloy is ordered.

Comparing the atomic structure of the system during heating (Fig. 2.3) and cooling (Fig. 2.4), i.e., in the course of the order–disorder and disorder–order transitions, it is easy to see the differences. Upon heating, the ordered regions maintain their ordering to higher temperatures in contrast to the case where ordered regions appear as a result of cooling.

Structural-phase transformations during the order–disorder and disorder–order phase transitions are of undoubted interest. Figures 2.5 and 2.6 show the change in the domain (phase) structure of the alloy as a function of temperature during heating (Fig. 2.5) and cooling (Fig. 2.6).

During heating to $T = 1000$ K, the alloy is completely ordered; the first disordered regions begin to appear in it at temperatures in the

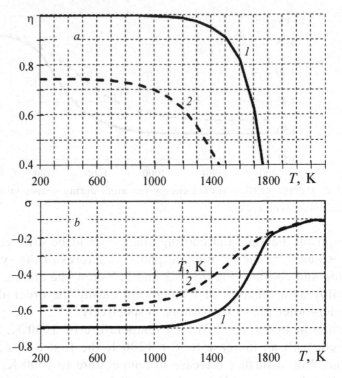

Fig. 2.2. Long-range (*a*) and short-range (*b*) order parameters in the process of sequential heating (curve 1) and cooling (curve 2).

range from 1100 to 1200 K. An increase in temperature to 1500 K leads to an increase in the number and size of disordered regions, and at $T = 1600$ K the disordered regions are already uniformly distributed throughout the system. With a further increase in temperature to 1800 K, almost the entire alloy is disordered, and at temperatures above 1900 K only domain nuclei remain in the alloy.

During the cooling stage, when the temperature decreases to 1800 K, only domain nuclei appear, and the sizes of the antiphase domains with the *B*2 superstructure increase during cooling to 1600 K. For a lower temperature (1300 K), no disordered regions remain in the alloy, except for the domain boundaries. It is significant that, with a further decrease in temperature, the shape of the boundaries changes, domains of the same type join, which persists up to 200 K. The difference in the average configuration energy per atom after the completion of the heating–cooling cycle (Fig. 2.1) is due to the formation of two antiphase domains during the disorder–order phase transition.

Comparing the phase distributions during heating (Fig. 2.5) and cooling (Fig. 2.6), we can conclude that the phase transition temperature during heating is higher than that during cooling, which is consistent

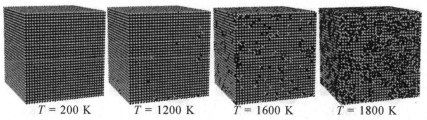

$T = 200$ K $\qquad$ $T = 1200$ K $\qquad$ $T = 1600$ K $\qquad$ $T = 1800$ K

Fig. 2.3. Temperature dependence of the atomic structure in stepwise heating.

with the temperature dependences of energy (see Fig. 2.1) and ordering (see Fig. 2.2). Comparison of the phase distribution during heating (order–disorder transition) and cooling (disorder–order transition) has led to the conclusion that the temperature of the phase transition during heating is higher than during cooling. This means that for the order–disorder transition to occur, the system needs to be slightly overheated relative to the traditionally understood phase transformation temperature, and for the disorder–order transition to occur, the system needs to be slightly supercooled relative to the same temperature. This is consistent with the temperature dependences of energy and ordering. After the completion of the disorder–order phase transition, two antiphase domains of the $B2$ superstructure are formed in the system.

Conclusion. Using the Monte Carlo method, an irreversibility of the processes has been shown which take place during thermal cycling in the course of the structural-phase transformations in BCC alloys, such as NiAl intermetallic compounds of the Ni–Al system. As a result of the heating and cooling cycle, a kind of hysteresis is observed in the curve, which proves the irreversibility of the processes, which implies a difference in the structural phase states at the heating and cooling stages. Analysis of the atomic and phase structure of the system in the course of heating and cooling, i.e. during the order–disorder and disorder–order phase transitions, confirmed the difference in structural phase states at the heating and cooling stages.

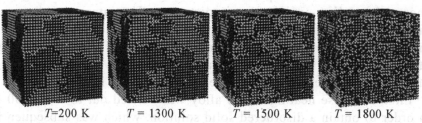

$T=200$ K $\qquad$ $T = 1300$ K $\qquad$ $T = 1500$ K $\qquad$ $T = 1800$ K

Fig. 2.4. Temperature dependence of the atomic structure in stepwise cooling.

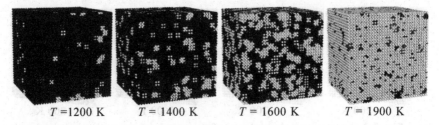

| $T = 1200$ K | $T = 1400$ K | $T = 1600$ K | $T = 1900$ K |

Fig. 2.5. Distribution of atoms in ordered and disordered phases in the NiAl alloy during heating.

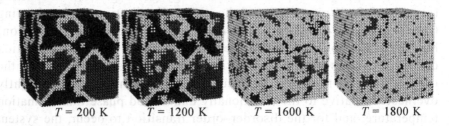

| $T = 200$ K | $T = 1200$ K | $T = 1600$ K | $T = 1800$ K |

Fig. 2.6. Distribution of atoms in ordered and disordered phases in the NiAl alloy during cooling.

Thus, for the order–disorder transition to occur, the system needs to be slightly overheated with respect to the traditionally understood phase transformation temperature, and for the order–disorder transition, the system needs to be slightly supercooled relative to the same temperature.

2.3. Features of the formation of antiphase domains in a NiAl alloy during stepwise cooling

Monte Carlo simulations show that the structural-phase characteristics of the BCC NiAl alloy are formed during cooling from a solid solution. The interaction between different pairs of atoms in this study is prescribed by the Morse pair potentials, the parameters of which are given in Table 2.1. The potential values are tabulated as changes in energy depending on interatomic distances. The interaction between different pairs of atoms is taken into account at a distance of the first three coordination spheres. Calculation of the configurational energy of the system, the short- and long-range orders, as well as other details of the modeling technique are given in section 1.1 in chapter 1.

First, stepwise heating of the alloy was carried out to $T = 2200$ K in order to obtain a disordered solid solution, which was subsequently subjected to stepwise cooling. In the process of stepwise cooling of the alloy, the mean configuration energy per atom (Fig. 2.7), the long-range η (Fig. 2.8 *a*) and the short-range σ order parameters (Fig. 2.8, *b*) were studied.

With a decrease in temperature to 2000 K, no changes in the energy values are observed. Further cooling leads to a smooth decrease in the energy value. However, the position of the curve during cooling differs from that during heating, which indicates the irreversibility of the heating–cooling process. It can be expected that the structural phase states upon heating would differ from those upon cooling. Comparing the heating and cooling curves, it is easy to see that the dependence curve $E = E(T)$ of cooling lies above the corresponding heating curve. It follows that the system has a certain specific energy value during cooling at significantly lower temperatures compared to the heating process. Thus, in order to achieve a certain energy value during the cooling process, it is necessary to supercool the system as compared to the heating process.

A sharp decrease in the configurational energy value corresponds to the temperature range from 1800 to 1300 K. It is likely that in this temperature range the low-stability states of the system are realized in the vicinity of the disorder–order phase transition. Taking into account $\sigma = \sigma(T)$ and $\eta = \eta(T)$ (Fig. 2.8), it is easy to conclude that it is in this temperature range that a hypothetical disorder–order transition occurs. It can be assumed that, in the case under consideration, in order to realize the order–disorder–order transition, the system must be superheated during heating to the temperature T_{0+}, and when cooled, it should be supercooled to T_{0-} with respect to the temperature of the reversible order–disorder–order transition.

From 1200 to 1000 K, the energy gradually decreases, i.e. after the disorder–order transition, the system lowers its energy only due to the ongoing ordering processes (Fig. 2.8). With further cooling below 900 K, the energy does not change.

The temperature range of variation of the short-range and long-range order parameters is consistent with the temperature range of the configuration energy change (Fig. 2.8). Negative values of the short-range order parameter indicate ordering trends. However, if upon cooling in the range from 1800 to 1300 K, the ordering stimuli increased rapidly with decreasing temperature, while in the range from 1200 to 1000 K, the changes are weaker (Fig. 2.8 *b*). Similar features are demonstrated by the behaviour of the long-range order parameter (Fig. 2.8 *a*).

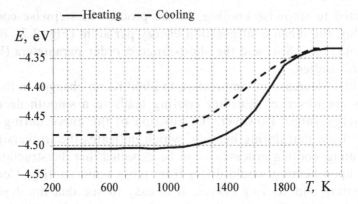

Fig. 2.7. Average configuration energy in the process of heating and stepwise cooling.

During cooling, there is no long-range order to $T \approx 1600$ K. With a decrease in temperature, a sharp increase in the values of the long-range order parameter is observed in the temperature range from ~1600 to ~1100 K. During heating, a rapid decrease in the values of the long-range order parameter is observed in the temperature range from ~1500 to ~1700 K. Judging by the curves $\eta = \eta(T)$ (Fig. 2.8 a) during heating and cooling, the order–disorder and disorder–order phase transitions occur in different temperature ranges, though no complete ordering of the system is attained during cooling. The value of the order parameter differs from unity at low temperatures. There may be several possible reasons. One of them is the formation of several antiphase domains during cooling of the system, which can lead to a decrease in the values of the long-range order parameter averaged over the entire system. Moreover, in each of the antiphase domains, the long-range order can be very high.

Sudden changes in the short-range order parameter during heating and cooling (Fig. 2.8 b) occur in the same temperature range of ~1100–1800 K. This indicates that the tendency to atomic ordering changes both in the heating and cooling stages in the same temperature range. It is in this very temperature range that significant changes in the long-range ordering occur (Fig. 2.8 a).

Let us dwell on the features of the formation of antiphase domains in the process of stepwise cooling. Figures 2.9 and 2.10 show the antiphase domains formed in the NiAl alloy during the disorder–order phase transition. Let us analyze the atomic (Fig. 2.9) and phase (Fig. 2.10) structures of the system during stepwise cooling, i.e., during the disorder–order phase transition. When the system is cooled (Fig. 2.9) with a decrease in temperature to 1800 K, regions ordered in accordance with the $B2$ superstructure appear. At $T = 1600$ K, the

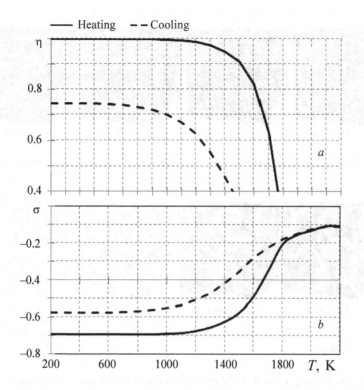

Fig. 2.8. Long-range (*a*) and short-range (*b*) order parameters during heating and stepwise cooling.

number and size of ordered regions increases, and at temperatures below 1300 K, the alloy is ordered.

Structural-phase transformations during cooling are of undoubted interest. Figure 2.10 presents the variations in the domain (phase) structure of the alloy as a function of temperature during cooling.

With a decrease in temperature to 1800 K, only domain nuclei appear, and the sizes of antiphase domains with the *B*2 superstructure increase upon cooling to 1600 K. The temperature of 1800 K is characterized by the appearance of isolated domain nuclei of various orientations in the alloy. Lowering the temperature to 1700 K leads to their growth and merging of domains of the same orientation. It is in the temperature range from 1800 to 1300 K that the most active processes of the formation of the domain structure are observed.

At 1500 K, a domain begins to transform into a structure corresponding to a two-domain crystal with the ordering elements along the boundaries. There are elements corresponding to superparticle dislocations in the <100> plane.

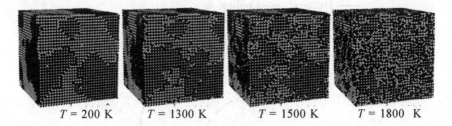

Fig. 2.9. Atomic structure of the alloy as a function of temperature in stepwise cooling.

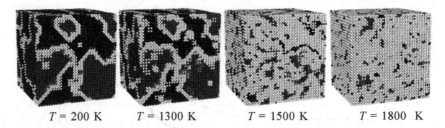

Fig. 2.10. Distribution of atoms in ordered and disordered phases in the NiAl alloy during cooling.

The ordering process develops with a decrease in temperature to 1400 K, while the regions corresponding to superparticle dislocations and antiphase boundaries are preserved.

Cooling to 1300 K preserves the elements of superdislocations and superdislocation loops, and the boundaries are smoothed out.

Lowering the temperature to 1200 K leads to an increase in the fraction of the ordered phase, the volume of the domain increases due to the attachment of the ordered near-boundary regions. At 1100 K, the process develops, the structure becomes more ideal, superparticle dislocations and APB disappear. The boundaries have predominantly <110> orientations.

At $T = 1000$ K, a boundary is formed between the domains clearly oriented along the <110> and <100> directions.

For a lower temperature, no disordered regions remain in the alloy, except for domain boundaries. It is significant that, with a further decrease in temperature (1300 K), the shape of the boundaries changes, domains of the same type join, which is maintained up to 200 K. The difference in the mean configuration energy per atom after completion of the heating–cooling cycle (Fig. 2.7) is due to the formation of two antiphase domains in the course of disorder–order phase transition.

A further decrease in temperature does not lead to any visible changes in the domain structure of the alloy.

Conclusion. Using the Monte Carlo method, it has been shown that two antiphase domains of the $B2$ superstructure are formed in the NiAl intermetallic compound after the completion of the disorder–order phase transition. In the course of domain formation, the elements corresponding to superparticle dislocations and superdislocation loops are observed. The domains are separated by antiphase boundaries in the <110> and <100> directions.

2.4. Effect of vacancy concentration on low-stability pre-transitional structural-phase states of NiAl No. 3 intermetallic compound

In this section we study the structural-phase features of low-stability pre-transitional states and the energy characteristics of intermetallic BCC compounds. by the example of NiAl intermetallic during heating and cooling, as a function of the concentration of point defects (e.g., vacancies) in the system. To do so, we consider the structural-phase states and energy characteristics of a defect-free NiAl intermetallic compound at $T = 0$ K. A defect-free system will subsequently act as the initial state. The lattice parameters of the starting configuration of the alloy are determined by the minimum configurational energy, which is derived by the gradient descent method. Next, we study the change in low-stability structural phase states and configurational energies of the NiAl intermetallic depending on the number of vacancies (1, 2, 5), provided that stoichiometry is preserved, which corresponds to a change in their concentration from $\sim 1.5 \cdot 10^{-5}$ to $\sim 7.6 \cdot 10^{-5}$. To activate the diffusion process, one vacancy is introduced into the system at random, which corresponds to a concentration of $\sim 1.5 \cdot 10^{-5}$. Only the vacancy diffusion mechanism is considered. The dynamic or kinetic component is present only in jumps of atoms to vacant sites.

It is known that nickel monoaluminide NiAl melts in an ordered state [1]. Disorder nonetheless occurs with increasing temperature. It is believed that the disordering temperature of the NiAl intermetallic is higher than its melting temperature. For this reason, we consider the hypothetical order–disorder transition during heating and disorder–order transition during cooling to study the laws of complex atomic ordering–disordering in order to increase the structural stability and mechanical properties of heat-resistant alloys based on the β-phase of the Ni–Al system and to determine the effect of disordering taking place with increasing temperature on the properties of the intermetallic compound.

In the course of modelling the system and the processes under consideration, we will follow the methodology described in section 1.1 in chapters 1.

In describing the interatomic interaction, we use the parameters of the Morse potentials given in Table 2.1.

Let us consider hypothetical order–disorder transitions during heating and disorder–order during cooling for the purpose of investigating the structural-phase features of low-stability pre-transitional states and the energy characteristics of the NiAl intermetallic during heating and cooling, depending on the concentration of vacancies in the system.

The curves of dependence of the average configurational energy on temperature during hypothetical order–disorder and disorder–order phase transitions are shown in Fig. 2.11 for a system with 1, 2, and 5 vacancies. It can be seen that during heating to a temperature of ~900 K, the energy practically does not change. In this temperature range, the effect of the concentration of vacancies on the average configurational energy of the system is not observed. In the range from ~1000 to ~1400 K, there is a gradual increase in the energy values.

In all cases under consideration, with an increase in the temperature of the system, the energy begins to increase from about 1100–1200 K. A certain difference arises in the range 1600–1800 K. Moreover, the more vacancies there are, the steeper the average energy dependence curve and the wider the temperature difference range. This can be understood in such a way that at a certain temperature in this temperature range, the structural-phase states of the systems with different vacancy concentrations differ. An increase in the number of vacancies increases the transformation velocity. The more vacancies in the alloy, the more intensively the processes of lowering the atomic order occur in it. This result is consistent with the laws obtained in [1] that an intensive increase in diffusion occurs due to an increase in the concentration of structural vacancies. Since it was shown [17] that a decrease in the atomic order in intermetallic compounds is a consequence of structural transformations, an increase in the concentration of vacancies, as can be expected, leads to an intensification of structural-phase transformations in intermetallic compounds. A sharp increase occurs in the range from 1400 to 1900 K. Apparently, this corresponds to a hypothetical order–disorder phase transition. At a temperature of ~1900 K, energy stops to increase.

During cooling, the situation is radically different. Upon cooling to ~1800 K, the energy gradually decreases. A decrease in temperature to below ~1600 K leads to differences in the temperature dependences of the average energy of the systems with different numbers of vacancies. This is likely to indicate different structural-phase states of the systems with a different number of vacancies. A sharp decrease in the configuration energy occurs in the temperature range from ~1800 to

~1300 K. At temperatures below ~1600 K, there is a strong dependence of the average configurational energy on the concentration of vacancies, and the higher the concentration, the closer this curve to the heating curve. This behaviour can be understood as follows: the more vacancies in the system at a certain temperature, the more intensive the diffusion processes, the faster the processes of increasing the atomic order, the lower would be the average energy of the system.

The difference in the average energy values after the completion of the transformation during the heating–cooling cycle (as a result of hypothetical order–disorder and disorder–order phase transitions) is due to the formation of two antiphase domains as a result of phase transitions in the system. Antiphase domains differ by their vacancy concentrations. This indicates a higher rate of atomic disordering or ordering in an alloy with a larger number of vacancies. Apparently, this is due to the intensification of diffusion processes with an increase in the defectiveness of the system due to an increase in the concentration of vacancies. As a result, this indicates a greater intensification of structural-phase transformations in a system with a larger number of defects in the intermetallic compound.

Note that during heating and cooling, the alloy undergoes various structural-phase states, which follows from the different positions of the heating and cooling curves of the system with a certain vacancy concentration. In turn, this indicates the irreversibility of diffusion processes during heating and cooling. Moreover, in the vicinity of the temperature $T \approx 1700$ K, the energies of various structural-phase states of the system differ very little, which suggests the realization of a whole range of different structural-phase states. This may imply

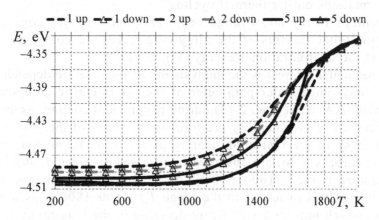

Fig. 2.11. Average configurational energy during heating and cooling with 1, 2 and 5 vacancies.

the formation of a certain set of pre-transitional low-stability states in a certain temperature range in the vicinity of the temperature of structural-phase transformations, which can be understood as the realization of a low-stability state of the system.

The behaviour of the short-range order parameter on the first coordination sphere of the NiAl intermetallic compound (Fig. 2.12) indicates the tendency of the alloy to ordering (a negative value of the short-range order parameter indicates a tendency to ordering). Naturally, with increasing temperature, the short-range order decreases.

During heating in the temperature range ~1600–1800 K, the values of the short-range order parameter differ for systems with different vacancy concentrations, which once again indicates a possible set of different structural-phase states in this temperature range, i.e. the realization of a low-stability state of the NiAl intermetallic in the pre-transitional temperature range.

Upon cooling, the structural phase states of systems with different vacancy concentrations differ even at the same temperature. The higher the defectiveness of the system (the higher the concentration of vacancies), the higher the temperatures in which the system is ordered due to the intensification of diffusion processes. This suggests that the higher the defectiveness of the structure, the higher would be the temperatures at which the structural-phase transformations occur in the intermetallic compound. From the behaviour of the curves of the temperature dependences of the short-range order parameter, it is easy to see the irreversibility of the processes of atomic disordering and ordering in the processes of heating and cooling of an intermetallic compound. This, in turn, suggests the irreversibility of structural phase transformations during thermal cycling.

Changes in the short-range order parameter for all considered configurational variants of the alloy (Fig. 2.12) are consistent with those in the configurational energy (Fig. 2.11).

Of particular interest is the analysis of the temperature dependences of the long-range order parameter of the intermetallic compound - monoaluminide NiAl (Fig. 2.13) during heating and cooling, especially in the region of low-stability pre-transitional states of the system.

It is easy to see that upon heating to $T \approx 1100$ K, no long-range order disturbances (η) are observed in all cases considered, and in the temperature range ~1100–1600 K, its value gradually decreases. A sharp decrease is observed in the range $T \approx 1600–1800$ K as η tends to zero, which implies a significant decrease in the long-range atomic order in the system. An increase in the number of defects, i.e., in the concentration of vacancies in the alloy, leads to a natural result – a decrease in the long-range order in the system in the region of low-stability pre-transitional states and a decrease in the transformation temperature.

Note that, at low temperatures, the concentration of vacancies does not significantly affect the degree of long-range order in the system nor the structural-phase state of the alloy. The effect of vacancy concentrations is manifested only in the region of pre-transitional low-stability states, and with an increase in the number of vacancies in the system, both the long-range order and the transformation temperature decrease significantly.

A very interesting scenario is realized when the intermetallide is cooled. There is no long-range order up to $T \approx 1600$ K. When the long-range order nucleates in the alloy, η substantially depends on the concentration of vacancies, i.e. the number of defects in the alloy. The greater the defectiveness, the more intense the diffusion processes are, and the faster the long-range order develops in the system grows. The more vacancies in the system, the higher the degree of long-range order at the same temperature. A sharp increase in the value of the long-range order parameter is observed when the temperature decreases from ~1600 to ~1200 K. With a further decrease in temperature, the steepness of the curve decreases to $T \approx 800$ K, and at lower temperatures the value of the long-range order parameter remains practically unchanged, but for each concentration has its own specific value .

The temperature range of variation of the short-range (Fig. 2.12) and long-range (Fig. 2.13) order parameters is consistent with the temperature range of the configurational energy (Fig. 2.11).

Conclusion. An analysis of the effect of the concentration of vacancies on the states and energy characteristics of the NiAl intermetallic compound during heating and cooling has demonstrated that the presence and concentration of vacancies turn out to be significant factors in the region of pre-transitional low-stability structural-phase states before the transformation. On the one hand, the presence of vacancies and their concentration do not affect the temperature ranges of structural phase transformations, while on the other hand, they significantly affect the pre-transitional low-stability structural phase states and the rate of diffusion processes.

It has been shown that during heating, an increase in the number of vacancies increases the transformation velocity. The more vacancies, the more intensively the processes of lowering the atomic order occur in the alloy. This results from the structural phase transformations; therefore, an increase in the concentration of vacancies, as can be

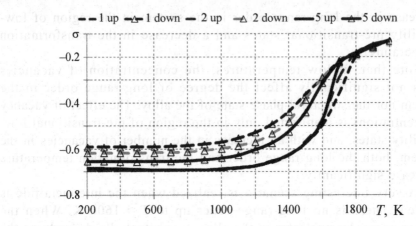

Fig. 2.12. Short-range order parameter in a sequential heating and cooling with 1, 2 and 5 vacancies.

expected, leads to an intensification of structural phase transformations in intermetallic compounds.

During cooling, a decrease in temperature leads to differences in the temperature dependences of the average energy of systems with different numbers of vacancies. The more vacancies in the system at a certain temperature, the more intensive the diffusion processes are, the faster the atomic ordering and the lower the average energy of the system. This occurs due to the intensification of diffusion processes with increasing defectiveness of the system due to an increase in the concentration of vacancies. As a result, this indicates a greater intensification of structural-phase transformations in a system with a higher concentration of vacancies, i.e., greater defectiveness.

It is noted that during heating and cooling the alloy undergoes various structural-phase states, which indicates the irreversibility of the processes during heating–cooling. Moreover, in the vicinity of the transformation temperature, the energy of the various structural phase states of the system hardlt differs, which suggests the formation of a number of different structural phase states. This, in turn, may indicate the realization of a certain set of pre-transitional low-stability states in a certain temperature range in the vicinity of the temperature of structural-phase transformations, which can be understood as the realization of a low-stability state of the system.

From the temperature behaviour of the short-range order parameter, it follows that the higher the defectiveness of the system, i.e. the higher the concentration of vacancies, the higher the temperature at which a tendency would develop for increasing the atomic order by virtue of diffusion processes. This, in turn, indicates an increase in the temperature of the onset of structural transformations with an increase

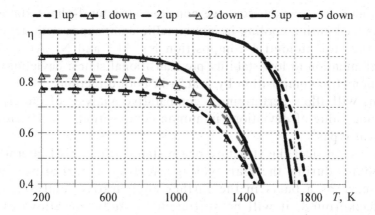

Fig. 2.13. Long-range order parameter dependence during sequential heating and cooling.

in the number of defects during cooling. From the behaviour of the temperature curves of the short-range order parameter, it is easy to see the irreversibility of the processes of decreasing the atomic order and its increase during heating and cooling of the intermetallic compound, which suggests the irreversibility of structural-phase transformations of the intermetallic compound during thermal cycling.

An analysis of the temperature dependences of the long-range order parameter of the intermetallic compound allows us to conclude that an increase in defectiveness, i.e. the concentration of vacancies in the alloy, leads to a natural result – a decrease in the long-range order in the system in the region of low-stability pre-transitional states and an increase in the temperature of the onset of transformation.

2.5. The effect of variations in atomic composition on low-stability pre-transitional structure-phase states of the NiAl intermetallic

Let us address the structure-phase features of low-stability pre-transitional states of intermetallic BCC compounds by the example of intermetallic compounds of the Ni–Al system as a functon of the atomic composition of the alloy: $Ni_{45}Al_{55}$, NiAl, $Ni_{55}Al_{45}$. The base for further analysis is the stoichiometric NiAl intermetallic compound during thermal cycling (hypothetical order–disorder transitions during heating and disorder – order transitions during cooling). This will allow us to establish the laws of complex atomic ordering and disordering, to determine the temperature range of low-stability pre-transitional states, and to reveal the effect of a decrease in the atomic order at elevated temperatures on the structure-phase state of the intermetallic compound. For comparison, we consider structural-phase states that are realized during cooling in intermetallic compounds with deviations from the stoichiometric composition $Ni_{45}Al_{55}$ and $Ni_{55}Al_{45}$, paying special attention to low-stability pre-transitional states: deviations from stoichiometric composition by 5 at.% in both directions.

This will allow performing a comparative analysis of the effect of the same value of deviations in atomic composition, for the atoms of different types.

Let us consider the structure-phase states of the NiAl intermetallic compound during thermal cycling (heating and cooling), which will serve as a basis in the further comparative analysis. Given this background, it will be sufficient to study the alloys of non-stoichiometric composition only during cooling. The starting configuration of the alloy will be set by a random distribution of Ni and Al atoms over the nodes of the BCC lattice in accordance with the given concentrations of the components. To activate the diffusion process, one vacancy is randomly introduced into the system, which corresponds to a concentration of $\sim 1.5 \cdot 10^{-5}$. Only the vacancy diffusion mechanism is considered. The dynamic or kinetic component is present only in jumps of atoms to vacant sites.

It is known that nickel monoaluminide NiAl melts in an ordered state [1], though its disordering occurs with increasing temperature. For this reason, let us consider hypothetical order–disorder transitions in the NiAl intermetallic compound during heating and disorder–order–transitions during cooling for the sake of investigating the laws of its complex atomic ordering–disordering and revealing the effect of disordering with increasing temperature on the properties of the intermetallic compound. Next, we consider in intermetallic compounds with deviations from the stoichiometric composition $Ni_{45}Al_{55}$ and $Ni_{55}Al_{45}$, paying particular attention to low-stability pre-transitional states. The temperature range of such states is found from an analysis of the order–disorder transitions (during heating) and the disorder–order transitions (upon cooling) in the NiAl intermetallic.

In modelling the system and the processes under consideration, we will follow the methodology described in Section 1.1 in Chapter 1.

In describing the interatomic interaction, we use the Morse potential parameters given in Table 2.1.

The curves of the temperature dependences of the average configurational energy during a hypothetical disorder–order phase transition are shown for three alloys: $Ni_{45}Al_{55}$, NiAl, and $Ni_{55}Al_{45}$ (Fig. 2.14).

From Fig. 2.14 it is easy to see that in the stoichiometric NiAl intermetallic compound at temperatures below 900 K, the energy does not change either upon heating or upon cooling, thourgh during its cooling is noticeably higher. A gradual increase in energy in the range from 1000 to 1400 K with an increase in temperature is replaced by a sharp increase in the range from 1400 to 1900 K, which corresponds, as can be assumed, to the realization of disordering processes (Fig. 2.15), i.e., indicating a hypothetical order–disorder phase transition occurs. During cooling, the energy values are higher, which is natural. A logical conclusion follows: to achieve atomic ordering in an alloy and to increase it, the alloy has to be overcooled. The difference between the heating and cooling curves means the irreversibility of diffusion processes during thermal cycling.

In non-stoichiometric alloys, the situation is somewhat different. As can be seen from Fig. 2.14, the curve of the configurational energy $E = E(T)$ of the $Ni_{55}Al_{45}$ alloy lies below the corresponding curves of the NiAl and $Ni_{45}Al_{55}$ alloys. The $Ni_{45}Al_{55}$ alloy curve lies above the corresponding NiAl and $Ni_{55}Al_{45}$ alloy curves. The dependence curves of stoichiometric alloys and nickel-enriched alloys are close both in position and in shape. However, the stoichiometric alloy curve at high temperatures has a steeper slope, i.e., the diffusion processes of atomic ordering occur more intensively; order in the alloy appears sooner and at a higher temperature. In the process of thermal cycling (heating-cooling) of the equiatomic alloy of NiAl, the features of structural-phase low-stability pre-transitional states were studied, paying attention, first of all, to the temperature behaviour of the average configurational energy per atom (Fig. 2.14), the short-range (Fig. 2.15, a) and long-range (Fig. 2.15, b) order parameters. It is easy to see that the temperature range of their variation (Fig. 2.15) is consistent with the temperature range of variation of configurational energy (Fig. 2.14). In the NiAl intermetallic of a stoichiometric composition, as a result of the heating and cooling cycles, a peculiar hysteresis is observed (Fig. 2.14), the presence of which during thermal cycling indicates the irreversibility of the processes and suggests a difference in the structural phase states at the heating and cooling stages. This is reflected in the behaviour of the system-average characteristics with temperature: the short-range (Fig. 2.15 a) and long-range (Fig. 2.15 b) orders parameters. Note that the numerical values of these characteristics differ significantly at the same temperature in the heating and cooling stages. Thus, even the behaviour of the average characteristics of the system indicates the irreversibility of the processes that occur during heating

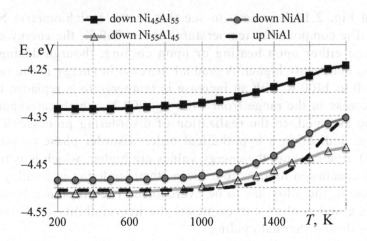

Fig. 2.14. Temperature dependence of the average configurational energy of intermetallic compounds during thermal cycling of NiAl and cooling $Ni_{45}Al_{55}$, $Ni_{55}Al_{45}$.

(hypothetical order–disorder phase transition) and cooling (hypothetical disorder–order phase transition) during thermal cycling.

For a comparative analysis Fig. 2.15 shows the changes in the short-range (Fig. 2.15 a) and long-range (Fig. 2.15 b) orders of the parameters of the three alloys differing in the composition of the considered intermetallic compounds during cooling. It is easy to see that the values of the short-range order parameter in all alloys are negative, and the negative value of the short-range order parameter indicates the tendency of the alloys towards ordering, with the equiatomic alloy being the most prone to ordering. The values of short-range order parameters in the alloys deviating from stoichiometry are close to each other during cooling.

The differences appear only at temperatures below ~1300 K. After step cooling, the parameter values in the $Ni_{55}Al_{45}$ and $Ni_{45}Al_5$ alloys are −0.47 and −0.44, respectively, and the short-range order parameter in the alloy with stoichiometric composition NiAl takes the value −0.59. Since the values of the short-range order parameter of non-stoichiometric alloys are substantially lower in their absolute value than those of the stoichiometric intermetallic compound, the tendencies to the atomic ordering in non-stoichiometric intermetallic compounds are significantly weaker than in the stoichiometric intermetallic compound. It was shown [17] that a decrease in the atomic order in intermetallic compounds is a consequence of structural transformations, so a deviation of the composition from the stoichiometric one, as can be expected from the temperature behaviour of the short-range order parameter (Fig. 2.15

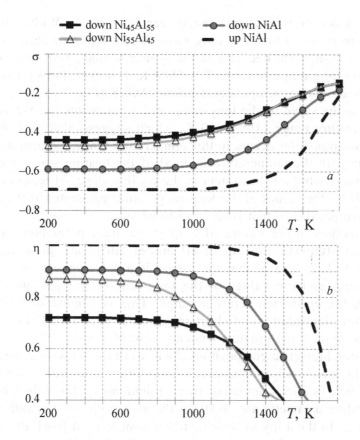

Fig. 2.15. Temperature dependence of the short-range (*a*) and long-range (*b*) orders of parameters on the concentration of alloy components during cooling.

a), will lead to a decrease in the intensification of diffusion processes during cooling: and will cause the need for supercooling and lowering the temperature of structural-phase transformations in intermetallic compounds of non-stoichiometric compositions.

The changes in the short-range order parameter for the alloys of various compositions (Fig. 2.15 *a*) are consistent with changes in configurational energy (Fig. 2.14).

During heating of the NiAl intermetallic of stoichiometric composition, as can be seen from Fig. 2.15 *b*, there are no disturbances of the long-range order of the order–disorder phase transition, up to a temperature of 1200 K, and then the long-range order values gradually decrease to a temperature of 1600 K. A sharp jump in the curve of the dependence of the long-range order parameter on temperature occurs at $T \approx 1700$ K, which indicates a rapid disordering in the system. It can be assumed that upon heating, the order–disorder phase transition occurs in a certain temperature range near $T \approx 1700$ K. With a further increase in temperature, the long-range order tends to zero, which indicates a significant decrease in the system's order. During cooling, there is no long-range order to $T \approx 1600$ K, however, the short-range order appears even at temperatures below 1800 K (Fig. 2.15 a). A sharp increase in the value of the long-range order parameter is observed in the temperature range 1100–1400 K, which suggests that a disorder–order phase transition occurs in this temperature range. Comparing this behaviour with the temperature interval of the heating stage, we can state an overcooling of the system in the cooling stage. At temperatures below 800 K, the values of the long-range order parameter in the alloy do not change; the system is in an ordered state.

The changes in the values of long-range order parameters in the process of stepwise cooling in all model alloys are presented in Fig. 2.15 b. It is logical that the long-range order appears at higher temperatures and the values of the long-range order parameter are higher in the entire temperature range. This is consistent with the temperature dependences of short-range order parameters (Fig. 2.15 a), which indicate the greatest tendency toward atomic ordering of an alloy of stoichiometric composition. As a result of stepwise cooling of the stoichiometric alloy, the value of the long-range order parameter reaches ~0.9. In alloys deviating from stoichiometry, the values of the corresponding parameter are ~0.89 for the alloy with $Ni_{55}Al_{45}$ and ~0.72 for $Ni_{45}Al_{55}$. In the alloy of stoichiometric composition NiAl, the value of the long-range order parameter increases faster during cooling than in the non-stoichiometric alloys.

The behaviour of the $\eta = \eta(T)$ curves during cooling of the alloys of non-stoichiometric compositions differs significantly from the corresponding behaviour of the curve of the stoichiometric alloy (Fig. 2.15 b). During cooling of the alloys of non-stoichiometric compositions, a substantial supercooling is necessary for the long-range order to be established, and the ordered phases appear at much lower temperatures. Moreover, the $\eta = \eta(T)$ dependence curve of the $Ni_{45}Al_{55}$ alloy lies significantly lower than the corresponding curve of the $Ni_{55}Al_{45}$ alloy. This, in turn, implies that the long-range order in these non-stoichiometric alloys is established in different ways, which is consistent with the available experimental data [1].

Attention should be paid to the behaviour of the $\eta = \eta(T)$ curves of the non-stoichiometric alloys in the temperature range from ~1200 to ~1500 K. In this range, the curve of the $Ni_{55}Al_{45}$ alloy lies below the corresponding curve of the $Ni_{45}Al_{55}$ alloy. At lower temperatures, the curves change their relative positions: the $Ni_{55}Al_{45}$ alloy curve lies already above the corresponding $Ni_{45}Al_{55}$ alloy curve. This implies that in the temperature range from ~1200 to ~1500 K, special structural phase states are realized. Only in this interval do the values of the long-range order of the $Ni_{45}Al_{55}$ alloy (or a phase in the heterophase mixture) exceed the corresponding values of the $Ni_{55}Al_{45}$ alloy. It would seem that, in this interval, a phase of stoichiometric composition can be decomposed into non-stoichiometric phases, with the overall atomic composition of the system being maintained. However, an analysis of the temperature dependences of the average configurational energy of the phases under consideration at various component concentrations (Fig. 2.14) during cooling indicates that this is not favorable from the point of view of the average configurational energy. In a multiphase system and in a non-equilibrium system this is quite possible.

Note the features that manifest themselves on non-stoichiometric alloys durin cooling (Fig. 2.15). Their deviation from the stoichiometric composition is 5% in both cases, however, the behaviour of the $\sigma = \sigma(T)$ and $\eta = \eta(T)$ curves is somewhat unusual. The behaviour of the $\sigma = \sigma(T)$ curves is similar (Fig. 2.15 a), but differs significantly from the behaviour of the $\sigma = \sigma(T)$ curve of a stoichiometric alloy. In non-stoichiometric alloys, the tendency to ordering is significantly weaker compared to the one in the stoichiometric alloy, as evidenced by the significantly lower σ values of the $Ni_{55}Al_{45}$ and $Ni_{45}Al_{55}$ alloys in absolute value.

It was shown [17] that a decrease in the atomic order in intermetallic compounds is a consequence of structural transformations, a deviation of the composition from the stoichiometric one, as can be expected from the temperature behaviour of the long-range order parameter (Fig. 2.15 b), will lead to a decrease in the intensification of diffusion processes in during cooling, for supercooling and lowering the temperature of structure-phase transformations in intermetallic compounds of non-stoichiometric compositions.

The temperature ranges of variations in the short-range (Fig. 2.15 a) and long-range (Fig. 2.15 b) order parameters in alloys with a deviation from stoichiometric composition are also consisten with the temperature range of changes in configurational energy (Fig. 2.14).

In the alloy of stoichiometric composition, NiAl, as a result of the heating and cooling cycle, a peculiar hysteresis is observed (Fig. 2.14), the presence of which during thermal cycling indicates the irreversibility of the processes, which suggests a difference in the structure-phase states within the heating and cooling stages. To this end, we analyze the atomic and phase structures of the system during heating and cooling, i.e., in the processes of hypothetical order–disorder and disorder–order phase transitions.

A qualitative analysis of the atomic structure of the NiAl alloy as a function of temperature during the order–disorder and disorder–order phase transitions confirmed the previous conclusions relying on the laws of the temperature behaviour of configurational energy and short-range and long-range order parameters. When heated to a temperature of ~1000 K, the alloy is ordered, with an increase of the temperature to $T = 1200$ K, the first disordered regions appear. With a further increase in temperature to 1600 K, the number and size of regions with disturbances of the superstructural arrangement of atoms increase. At temperatures above 1800 K, the superstructural arrangement of the atoms is substantially disturbed, and when the temperature reaches 1900 K, the alloy is completely disordered. As the system cools with decreasing temperature to ~1800 K, regions appear that are ordered in accordance with the $B2$ superstructure. At $T \approx 1600$ K, the number and size of ordered regions increases, and at temperatures below 1300 K, the alloy is ordered. Comparing the atomic structure of the system in heating and cooling, i.e., in the course of hypothetical order–disorder and disorder–order phase transitions, it is easy to see the differences. Upon heating, the ordered regions are retained to higher temperatures in comparison with the temperatures of the appearance of ordered regions upon cooling.

Structure-phase transformations during hypothetical order–disorder and disorder–order phase transitions are of undoubted interest. Figure 2.16 shows the change in the domain (phase) structure of the stoichiometric NiAl alloy depending on temperature during heating (Fig. 2.16 a) and cooling (Fig. 2.16 b).

Dark areas indicate the regions ordered in accordance with the $B2$ superstructure, light areas correspond to disordered structures and grain boundaries. During heating to $T \approx 1000$ K, the alloy is completely ordered; at temperatures in the range from 1100 to 1200 K, the first disordered regions begin to appear in the alloy. An increase in temperature to 1500 K leads to an increase in the number and size of disordered regions, and at $T \approx 1600$ K, disordered regions are already uniformly distributed throughout the system. With a further increase in temperature to 1800 K, almost the entire alloy is disordered, and at temperatures above 1900 K only domain nuclei remain in the alloy.

For a comparative analysis, Fig. 2.17 shows the distribution of atoms over ordered and disordered phases during cooling. It is easy to see that upon cooling, the first ordered regions appear in the NiAl stoichiometric alloy at $T \approx 1600$ K, and these regions are significant in size. At this temperature, single ordered regions are observed in alloys of non-stoichiometric composition $Ni_{45}Al_{55}$ and $Ni_{55}Al_{45}$. With decreasing temperature (for example, at $T \approx 1400$ K), the ordered regions in the alloy of stoichiometric composition NiAl rapidly increase in size, while in alloys of non-stoichiometric composition $Ni_{45}Al_{55}$ and $Ni_{55}Al_{45}$, an increase in the number of ordered regions is observed. Therefore, for the appearance of ordered regions in alloys of non-stoichiometric composition, $Ni_{45}Al_{55}$ and $Ni_{55}Al_{45}$, additional supercooling is necessary in comparison with an alloy of stoichiometric composition NiAl. Moreover, from an analysis of the sizes of ordered regions, we can assume that atoms exceeding the stoichiometric ratio can serve as centers of nucleation of the ordered phase.

It should be noted that when the composition of the system deviates from the stoichiometric (Fig. 2.17), significant refinement of the ordered and disordered regions is observed. For example, at the same temperature $T \approx 1200$ K, the sizes of ordered and disordered regions in $Ni_{45}Al_{55}$ and $Ni_{55}Al_{45}$ intermetallic compounds are significantly inferior to the corresponding sizes of similar regions in NiAl intermetallic compounds. It can be assumed that this is due to the appearance of Ni or Al atoms, which cause the composition to deviate from stoichiometric. Local deviations are additional centres of local structural-phase transformations.

Comparing the sizes of the ordered and disordered regions in the non-stoichiometric intermetallic compounds $Ni_{45}Al_{55}$ and $Ni_{55}Al_{45}$, it can be noted that the temperature dependences of the regions of these alloys are different. It can be assumed that although deviations from the stoichiometric composition are of the same value, the nature of the ordering in them differs due to the difference in the atoms deviating from stoichiometry.

Conclusion. Using the Monte Carlo method, we have studied the effect of compositional variations on the characteristics of pre-transitional low-stability structure-phase states of the NiAl intermetallic in the region of structure-phase transformations. It has been shown that in the NiAl intermetallic compound of the stoichiometric composition, a peculiar hysteresis is observed during thermal cycling, the presence of which indicates the irreversibility of the processes taking place. This implies a difference in the structure-phase states in the stages of heating and cooling.

An analysis of the influence of the deviation of the atomic composition from the stoichiometric on the state of the NiAl intermetallic compound during cooling showed that the deviation turns out to be a significant factor in the region of pre-transitional

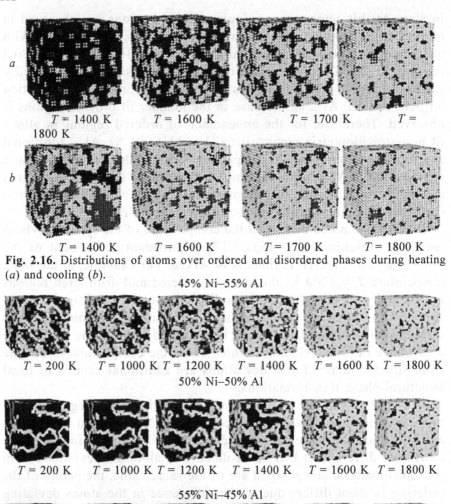

a

$T = 1400$ K $T = 1600$ K $T = 1700$ K $T =$
1800 K

b

$T = 1400$ K $T = 1600$ K $T = 1700$ K $T = 1800$ K

Fig. 2.16. Distributions of atoms over ordered and disordered phases during heating (*a*) and cooling (*b*).

45% Ni–55% Al

$T = 200$ K $T = 1000$ K $T = 1200$ K $T = 1400$ K $T = 1600$ K $T = 1800$ K

50% Ni–50% Al

$T = 200$ K $T = 1000$ K $T = 1200$ K $T = 1400$ K $T = 1600$ K $T = 1800$ K

55% Ni–45% Al

$T = 200$ K $T = 1000$ K $T = 1200$ K $T = 1400$ K $T = 1600$ K $T = 1800$ K

Fig. 2.17. Distributions of atoms over ordered and disordered phases in intermetallic compounds of the Ni–Al system during cooling.

low-stability structure-phase states before the transformation. The values of the short-range order parameter of the non-stoichiometric alloys are much smaller in absolute value of the corresponding values of the stoichiometric intermetallic compound; therefore, the tendencies to atomic ordering in non-stoichiometric intermetallic compounds are significantly weaker than the stoichiometric intermetallic compound.

Since it was previously shown that a decrease in the atomic order in intermetallic compounds is a consequence of structural transformations, a deviation of the composition from the stoichiometric one, as can be expected from the temperature behaviour of the short-range order parameter, will lead to a decrease in the intensification of diffusion processes during cooling, the need for supercooling, and a structural decrease in temperature of the structure–phase transformations in intermetallic compounds of non-stoichiometric compositions.

The behaviour of the temperature dependence of the long-range order parameter of alloys of non-stoichiometric compositions in a certain temperature range in the vicinity of the disorder–order transition has been given particular attention. In this range, the $Ni_{55}Al_{45}$ alloy curve lies below the corresponding $Ni_{45}Al_{55}$ alloy curve. At lower temperatures, the curves change their relative positions: the $Ni_{55}Al_{45}$ alloy curve lies already above the corresponding $Ni_{45}Al_{55}$ alloy curve. From this it can be assumed that in this temperature range special structure-phase states are realized. Only in this interval do the values of the long-range order of the $Ni_{45}Al_{55}$ alloy (or phase in the heterophase mixture) exceed the corresponding values of the $Ni_{55}Al_{45}$ alloy. It would seem that, in this interval, a phase of stoichiometric composition can be decomposed into phases of non-stoichiometric compositions, while maintaining the overall atomic composition of the system. However, an analysis of the temperature dependences of the average configurational energy of the phases under consideration at various concentrations of the components during cooling indicates that this is not favourable from the point of view of the average configurational energy. On the other hand, in a multiphase system and in a nonequilibrium system this is quite possible.

During cooling, the behaviour of the temperature dependences of the long-range order parameter of non-stoichiometric alloys differs significantly from the corresponding behaviour of the dependence of the alloy of stoichiometric compositions. To establish long-range order in alloys of non-stoichiometric compositions, significant supercooling is required, and the appearance of ordered phases occurs at lower temperatures. In addition, the temperature dependence of the long-range order parameter of the $Ni_{45}Al_{55}$ alloy lies substantially lower than the corresponding curve of the $Ni_{55}Al_{45}$ alloy. This, in turn, implies that the long-range order in these non-stoichiometric alloys is established in different ways.

It should be noted that when the composition of the system deviates from the stoichiometric one, significant refinement of ordered and disordered regions is observed.

2.6. Effect of grain size on low-stability pre-transitional structure-phase states of NiAl intermetallic compound

Using the Monte Carlo method, in this section we study the effect of the grain size (the size of the computational cell) of the alloy on the features of the pre-transitional low-stability structure-phase states of the NiAl intermetallic in the region of structure-phase transformations during cooling. To do this, we consider three model alloys having a different 'grain' size. Fine'-grain-' – a computational cell size of 16 × 16 × 16 atomic layers, i.e. 8192 atoms; the average is the size of the calculated block cell 32 × 32 × 32 atomic layers, i.e. 65 536 atoms; large – a computational cell size of 48 × 48 × 48 atomic layers, i.e. 221184 atoms. This will allow us to carry out a comparative analysis of the influence of the grain size of the alloy (the computational cell size) on the features of pre-transitional low-stability structural-phase states of the NiAl intermetallic in the region of structure-phase transformations.

We study the structure-phase features of low-stability pre-transitional states and the energy characteristics of the NiAl intermetallic during thermal cycling: heating (a hypothetical order–disorder transition) and cooling (a hypothetical disorder–order transition) depending on the grain size of the alloy (size of the calculated cell unit). In the initial state, all alloys were in an ordered state with a $B2$ superstructure.

In modelling the system and the processes under consideration, we will follow the methodology described in Section 1.1 in Chapter 1.

In describing the interatomic interaction, we use the parameters of the Morse potentials given in Table 2.1.

First, stepwise heating is carried out from 200 to 2000 K, then stepwise cooling to 200 K. The temperature step is 100 K.

The temperature dependence of configurational energy for alloys with different grain sizes is shown in Fig. 2.18. From the type of graphs, it can be assumed that during heating, the transformation (a hypothetical order–disorder transition) occurs in a certain temperature range: with fine grains from ≈1400 to ≈1700 K, with average grains from ≈1400 to ≈1900 K, with large grains from ≈1400 to ≈1900 K. When heated, the temperature range of the transformation is sensitive to grain size: as one can assume, with increasing grain size, the temperature range of the transformation would extend. During cooling, the transformation (a hypothetical disorder–order transition) occurs in a certain temperature range: with fine grains from ≈1500 to ≈1200 K, with medium grains from ≈1500 to ≈1000 K, with large grains from ≈1500 to ≈800 K. It follows from this that upon cooling, the transformation temperature range is also sensitive to grain size: with increasing grain size, the temperature range of the transformation becomes wider.

Thus, with an increase in grain size, the temperature range of the transformation increases both during heating and during cooling. It is easy to visually note that with an increase in the grain size of the alloy, the temperature range of the transformation becomes wider.

On the other hand, there is an increase in the temperature difference of the transformations during heating and cooling. This is especially pronounced during cooling: with fine grains, the transformation interval is ~300 K, with medium-size grains of ~500 K, and with large grains ~800 K.

It is easy to see hysteresis loops whose areas increase with increasing grain size. The presence of a hysteresis loop during thermal cycling indicates the irreversibility of the processes and suggests the difference in the structural phase states at the stages of heating and cooling. For an alloy with small grains (with a small number of atoms in the 'computaitonal cell' – 8192), the energy value after termination of the thermal cycle (successive order–disorder and disorder–order phase transitions) coincides with the initial value. It follows from this that in the case of fine grains the loop is closed, i.e. structure-phase states after the completion of the thermal cycle at temperatures below ≈ 1200 K correspond to structure-phase states at the same temperature until the thermal cycle is realized. The final state of the alloy structure corresponds to a fully ordered structure. With increasing grain, the loop is no longer closed, i.e., the processes and the differences in structural phase states during heating and cooling at the same temperature become irreversible. Therefore, the manifestation of the irreversibility of the processes and the difference in the structure-phase states during heating and cooling depend on the grain size of the alloy.

It is important to consider the laws of temperature dependences of the short-range (Fig. 2.19 a) and long-range (Fig. 2.19 b) order parameters for different grain sizes of the alloy.

Analyzing the laws of temperature dependences of short-range order parameters (σ) for different grain sizes of the alloy, it should be noted that all alloys have negative σ values. This indicates the tendency of the alloys under consideration to atomic ordering.

During heating to $T \approx 1300$ K, the σ values of all alloys coincide, however, at higher temperatures they begin to differ. The alloy with the largest grain size shows the strongest tendency for ordering, and in the case of fine grains this tendency is less pronounced.

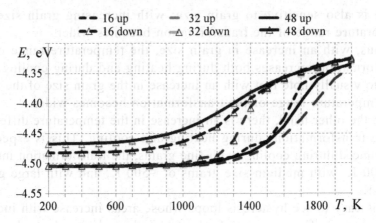

Fig. 2.18. Temperature dependence of configuration energy on the size of the computational cell of the NiAl alloy during thermal cycling.

Under cooling, the situation is reverse. The alloys with the smallest grain size are most prone to ordering, and with grain enlargement this tendency decreases.

While with a fine grain the loop of the dependence $\sigma = \sigma(T)$ is closed, with increasing grain size it is moving further away from the closed loop. It can be expected that the structure-phase states after the thermal cycle would increasingly differ from the structure-phase state to the thermal cycle at the same temperature.

It is easy to see that as the grain size increases, the temperature interval of the structure-phase transformations increases (the difference in the transformation temperatures during heating and cooling), i.e. loop area increase.

When analyzing the behaviour of the temperature dependences of long-range order parameters (η) for different grain sizes of the alloy (Fig. 2.19 *b*), it should be noted that upon heating all alloys have an almost complete atomic order to $T \approx 1300$ K. With a further increase in temperature, the dependences $\eta = \eta(T)$ begin to differ in the cases of different grain sizes. The most complete long-range order is observed in the alloy with the largest grain size, and the smallest – with the smallest grain size. To disorder the alloy with increasing grain size, a significant increase in temperature is required, i.e. significant overheating of the alloy. This is combined with the well-known fact about the melting of NiAl intermetallide with a large grain size in an ordered state [1].

Under cooling, the situation is somewhat different (Fig. 2.19 *b*). Long-range order appears, first of all, in an alloy with fine grains. As the grain size increases, the temperature of the appearance of long-range order decreases, i.e. with increasing grain size, more and more subcooling is required to realize the atomically ordered state of the system. The long-range order in the alloy with fine grain reaches ~1.0, while in the alloy with medium grain it is ~0.75, and in the alloy with large grain it is ~0.60. After the heating-cooling cycle is completed, the parameter value is different from 1, which indicates the formation of antiphase domains of the $B2$ superstructure.

It is easy to see that the value of the long-range order parameter of the alloy during stepwise heating depends on the grain size (computational cell size). The larger it is, the wider the temperature range of the structure-phase transformation (order–disorder and disorder–order phase transitions).

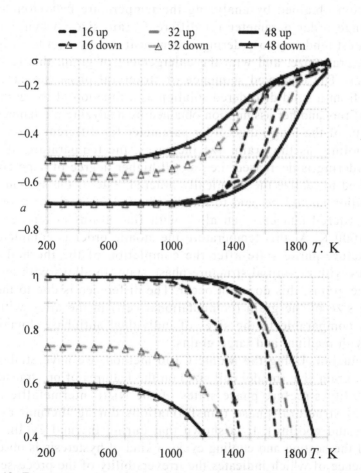

Fig. 2.19. Temperature dependence of the short-range (a) and long-range (b) order parameters for different grain sizes – the computational cell of the NiAl alloy in the process of thermal cycling.

The change in the values of the short- and long-range order parameter is consistent with changes in the average configurational energy of alloys with different grain sizes during thermal cycling (Fig. 2.18).

The features of the formation of structure-phase states during cooling from a solid disordered solution depending on the grain size (computational cell) are shown in Fig. 2.20. Dark regions correspond to regions ordered in accordance with the $B2$ superstructure, light regions correspond to disordered structures and grain boundaries.

From Fig. 2.20 it is easy to see that during cooling, the first ordered regions appear in the alloy with fine grain ($T \approx 1800$ K). In an alloy with medium grain, this happens at $T \approx 1600$ K, and in an alloy with large grain at $T \approx 1600$–1400 K. This regularity confirms the assumption obtained by analyzing the temperature behaviour of the short-range order parameter for different grain sizes: when cooling, the greatest tendency to ordering is demonstrated by an alloy with the smallest grain size, and with the enlargement of grain, this tendency decreases. The observed sequence of structural phase states during cooling from a solid disordered solution as a function of the grain size also confirms another assumption obtained by analyzing the temperature behaviour of the long-range order parameter for different grain sizes: upon cooling, as the grain size increases, the temperature of long-range ordering is decreased, i.e., a still higher degree of supercooling is required to realize the atomically ordered state of the system.

Attention should be paid to the distribution of atoms over ordered and disordered phases in an alloy with fine grains at temperatures below 1000 K. At this temperature the atomic order is complete, i.e. the structure-phase state after the completion of the thermal cycle coincides with the initial structure-phase state. In alloys with medium and large grains, this does not occur. The difference is due to the fact that the size of the grain (computational cell) in the alloy with fine grain is comparable to the sizes of individual antiphase domains in alloys with medium and large grains.

Conclusion. Using the Monte Carlo method, we have studied the effect of grain size (model cell size) on the features of pre-transitional low-stability structure-phase states of the NiAl intermetallic in the region of structural-phase transformations during thermal cycling (heating and cooling). It is shown that during thermal cycling as a result of the heating and cooling cycle, a kind of hysteresis is observed, the presence of which indicates the irreversibility of the processes that occur, which implies a difference in the structure-phase states at the heating and cooling stages. With increasing grain size, the areas of hysteresis loops increase. In the case of fine grains the loop is closed, i.e., at low temperatures, the structure-phase states after the completion of the thermal cycle correspond those at the same temperature before the thermal cycle. With increasing grain, the loop loses its closure, i.e. in the entire temperature range, the irreversibility of processes and the difference in the structure-phase states during heating and cooling at the same temperature begin to appear. It follows that the manifestation of the irreversibility of the processes and the difference in the structure-phase states during heating and cooling depend on the grain size of the alloy. With an increase in grain size, the temperature range of the transformation increases both during heating and cooling.

An analysis of changes in the short-range order parameter showed that in the region of low-stability pre-transitional states upon heating, the alloy with the largest grain size shows the greatest tendency to ordering, and this tendency decreases during grain refinement. During cooling, the alloy with the smallest grain size shows the greatest tendency toward ordering, and with grain enlargement this tendency decreases. As the grain size increases, the temperature range of the structural-phase transformations (the difference in the transformation temperatures during heating and cooling) is extended.

An analysis of the temperature dependences of the long-range order parameter revealed that upon heating, the most complete long-

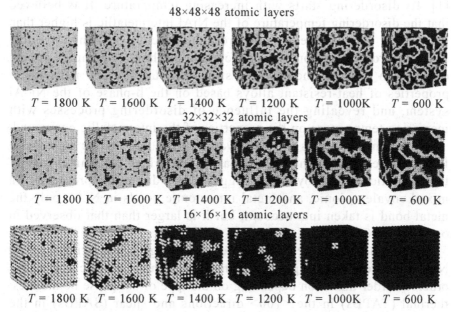

Fig. 2.20. Distributions of atoms by ordered and disordered phases in NiAl alloys with different grain sizes during cooling

range order is observed in the alloy with the largest grain size, and the smallest – with the smallest grain size. To disorder an alloy with a growing grain size, it is necessary to superheat the alloy. Upon cooling, the long-range order primarily appears, in the alloy with fine grain. As the grain size increases, the temperature of the appearance of long-range order decreases, i.e. with increasing grain size, a still higher subcooling is required to realize the atomically ordered state of the system. The larger the grain size, the wider the temperature range of the structure-phase transformation (order–disorder and disorder–order phase transitions).

The features of the formation of structure-phase states during cooling from a solid disordered solution as a function of the grain size (model cell size) show that the first ordered regions appear in the alloy with fine grain. As the grain size increases, the temperature of the appearance of long-range order decreases, i.e. with increasing grain size, a still higher subcooling is required to realize the atomically ordered state of the system.

2.7. Structure-phase low-stability states of BCC intermetallic compounds with APB complexes

It is known that nickel monoaluminide (NiAl) melts in an ordered state [1]. Its disordering starts with increasing temperature. It is believed that the disordering temperature of the NiAl intermetallic is higher than its melting temperature. For this reason, we consider a hypothetical order–disorder transition in order to study the laws of complex atomic ordering in order to increase the structural stability and mechanical properties of heat-resistant alloys based on the β-phase of the Ni–Al system, and revealing the influence of disordering processes with increasing temperature on the properties of the intermetallic compound.

We assume that it is the large value of the interatomic interaction forces in the lattice of nickel monoaluminide which determines mainly the properties of β-alloys. This suggests that, due to the presence of a mixed covalent, ionic, and metal interatomic bonds in NiAl, only the metal bond is taken into account, but it is larger than that observed in ordinary BCC alloys.

Let us study the structural features and energy characteristics of the NiAl intermetallic compound during a hypothetical structural-phase order–disorder transition depending on the type of antiphase boundaries: thermal (TAPBs) in the <100> directions and shear (SAPBs) in the <100> directions.

Consider an ordered BCC structure with the *B*2 superstructure (NiAl intermetallic compound) (see Fig. 1.1). Let the model (computational cell) include 32 × 32 × 32 unit cells (65536 atoms), and use periodic boundary conditions, which effectively corresponds to an infinite system with a long period.

The energy of formation of the antiphase boundary will be found as

$$E^* = \frac{E_{APB} - E}{S},$$

where E is the configurational energy of an ideal defect-free alloy; E_{APB} is the configurational energy of a system with APB; S is the APB area.

We study the change in low-stability structure-phase states and β-brass configuration energies during the order-disorder phase transition, short-range and long-range order parameters for three model configurations of the NiAl intermetallic compound: without APBs, with a complex of shear APBs in the <110> direction and with a complex of thermal APBs in the <100> direction. Complexes will be built from a pair of antiphase boundaries of the corresponding direction, taking into account the conservation of stoichiometry.

First, we consider a defect-free system (a defect-free NiAl intermetallide) during the order–disorder phase transition. In the future, this defect-free system will act as the initial state.

TAPBs normal to the <100> direction are considered. In this direction, in the *B*2 superstructure, the planes of nodes that are appropriate for Ni and Al atoms alternate (see Fig. 1.1). Thermal antiphase boundaries are set by subtracting the planes of Ni or Al atoms. Subtracting the plane of the Ni atoms, we obtain the thermal antiphase boundary and pairs of the nearest neighbours Al–Al (hereinafter, we will call this TAPB an Al–Al type boundary). Subtracting the plane of the Zn atoms, we obtain the thermal antiphase boundary and pairs of the nearest Ni–Ni neighbours (hereinafter, we will call this TAPB a Ni–Ni type boundary). The Al–Al and Ni–Ni type boundaries constitute a dual complex in which TAPBs are spaced at a certain distance. Note that with the introduction of such a dual complex, the equiatomic composition of the system does not change. In the presented images of the structural phase state of the system, the Zn–Zn boundary will be located on the right, and the Cu–Cu boundary will be on the left.

We draw attention to changes in the low-stability structural-phase state of the system as a result of the action of the vacancy diffusion mechanism during the order-disorder phase transition in a system with a pair of TAPBs (dual complex) separated by eight unit cells in the <100> direction.

In the direction <110> in the $B2$ superstructure only shear antiphase APBs can exist. To preserve the stoichiometry of the composition, the boundaries are introduced at an angle of 90° in planes passing through the centre of the computational block. Therefore, in the central part and along the boundaries of the computational block, we have the regions of intersection of the shear boundaries. Border crossing is a monoatomic column the size of one unit cell.

When finding the long-range order parameter [17]

$$\eta = \frac{P_A^{(1)} - C_A}{1 - v},$$

it is necessary to calculate the probability. To do this, we determine the number of atoms of sort A, in which the neighbors on the first sphere correspond to the first and second type of domains (parts of the system located on opposite sides of the antiphase boundary). Accordingly, nodes of the first type are considered to be all nodes that are legal for atoms of type A, depending on the type of domain:

$$P_A^{(1)} = \frac{N' + N''}{N_1},$$

where N_1 is the number of nodes of the first type; N' is the number of atoms of sort A in the nodes of the sublattice of the first type; N'' is the number of atoms of sort A located in the nodes of the other sublattice and ordered in accordance with the domain of the second type on the first sphere.

In describing the interatomic interaction, we use the parameters of the Morse potentials given in Table 2.1. The potential values were tabulated as changes in energy depending on interatomic distances.

The study will be carried out in the following sequence: we will consider the structure-phase states and energy characteristics of a defect-free CuZn alloy, an alloy with a complex of antiphase boundaries in the <100> direction (a pair of thermal APBs) and with a complex of antiphase boundaries in the <100> direction (a pair of shear APBs). After a comparative analysis, we reveal the structure-phase and energy features of the low-stability states of the NiAl intermetallide during the order–disorder phase transition.

Let us consider the NiAl intermetallide alloy in a defect-free structural state and in states with TAPB and SAPB complexes. The temperature dependence of the average configurational energy during the order-disorder phase transition is shown in Fig. 2.21.

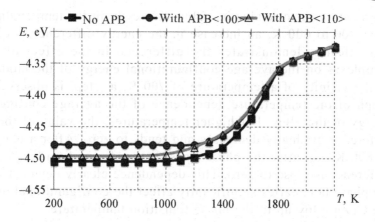

Fig. 2.21. Temperature dependence of the average configuration energy of NiAl intermetallide

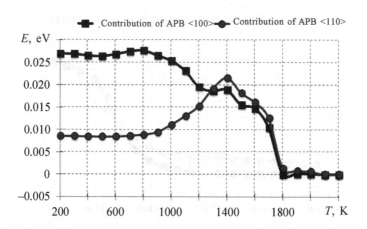

Fig. 2.22. Contribution of APB complexes to configuration energy with increasing temperature,

It is easy to see that, over the entire temperature range, the energies of states of an alloy with antiphase boundaries are higher than that of defect-free ones. Moreover, the values of the average configurational energy of an alloy with a complex of thermal APB in the <100> direction are significantly higher than those of an alloy with shear APB in the <110> direction.

In the range from 400 to 500 K, a decrease in the energy difference is observed, which indicates the influence of APBs on the structural–energy characteristics of the alloy during heating. The average configurational energy of the alloy with the SAPB complex at low temperatures also retains its value. However, in the temperature range from 400 to 600 K, an increase in the energy difference is observed. This clearly demonstrates the difference in the effect of various complexes on the average configurational energy of the model alloy. In the vicinity of a temperature of 700 K, a 'step' is observed in the graph of the temperature dependence of the average configurational energy of the alloy. For higher temperatures, the value of the energy difference gradually decreases and tends to zero. After a temperature of 800 K, i.e. after the phase transition, order is disorder, the energy difference is close to zero. This dependence clearly demonstrates that antiphase boundaries significantly affect the configurational energy of the CuZn alloy up to the phase transition temperature.

The behaviour of the short-range order parameter on the first coordination sphere, in particular its negative value (Fig. 2.23), indicates the tendency of the alloy to ordering (a negative value of the

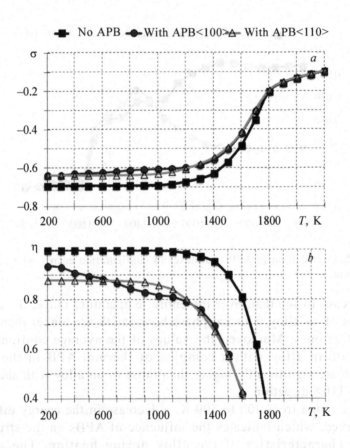

Fig. 2.23. Temperature dependence of short-tane (a) and long-range (b) order in the intermetallide - monoaluminide NiAl.

short-range order parameter indicates a desire for ordering) However, with increasing temperature, degree of short-range ordering decreases. A system with a complex of shear APBs is more prone to ordering than a system with a complex of thermal APBs. The variations in the short-range order parameter for all considered configurational variants of the alloy (Fig. 2.23 *a*) are consistent with the configurational energy.

Let us now address the temperature dependence of the long-range order parameter of the NiAl intermetallic compound in the region of low-stability pre-transitional states, for which the hypothetical order–disorder phase transition process is considered (Fig. 2.23 *b*).

It is clearly seen that in a defect-free system (defect-free model NiAl intermetallic compound) up to $T = 1100$ K, there are no long-range order η disturbances, and in the temperature range ~1100–1600 K its value gradually decreases. A sharp decrease is observed in the interval $T \approx 1600–1800$ K when the value of η tends to zero, which implies a significant disordering in the system. The introduction of the APBs in the alloy leads to a natural result – a decrease in the long-range order in the system in the region of low-stability pre-transitional states for both types of APBs (Fig. 2.23 *b*). Note that at low temperatures the value of the long-range order parameter in a system with a complex of TAPBs is higher than that in a system with a complex of SAPBs. However, at $T \approx 700$ K, the η values in both considered variants of the system with complexes of defects become equal, and with a further increase in temperature, the degree of the long-range order in the system with the SAPB complex becomes higher.

In the NiAl intermetallic compound with a complex of thermal APBs in the <100> direction, structural order disturbances begin at low temperatures (from 300 K), in the range from 900 to 1200 K, the long-range order parameter preserves its values, a further increase in temperature leads to a sharp decrease in its value. The configuration of a monoaluminide with a complex of shear antiphase atoms in the <110> direction, on the contrary, demonstrates a stable value of the long-range order parameter up to 1000 K, and when the temperature rises to about 1400 K, the value of the long-range order parameter gradually decreases. In a system with APB complexes, the long-range order parameter naturally decreases faster compared to a defect-free alloy.

It is worth noting that the positive values of the energy of the thermal APB complex in the <100> direction are larger than the values of the energy of the shear APB complex in the <110> direction. Deviations of the values of the short-range order parameter from the ideal value (a negative value of the short-range order parameter indicates a tendency towards for ordering) in an alloy with a complex of thermal APBs are also larger than those of the short-range order parameter in the case of a system with a complex of shear APBs. At the same time, at low temperatures, the deviations of the η parameter values of a system with a complex of thermal APBs are smaller than those of the η values of a system with a complex of shear APBs. In fact, there is some inversion of this dependence. This might be due to the fact that even at low temperatures, the regions of boundary intersection in the alloy with a complex of shear antiphase states transfer to an ordered state, which affects the long-range order. The most significant for the long-range order in the system is the appearance of a structural defect in the form of antiphase atoms, the difference in the plane of their occurrence does not strongly affect the behaviour of η with temperature. Naturally, a system with structural defects is less ordered than a defect-free system, which is manifested in the behaviour of the $\eta = f(T)$ curves: the curve of the defect-free intermetallic curve lies above the curves of the NiAl monoaluminide with APBs. The presence of a defect in the form of an APB promotes the onset of disordering of the system at lower temperatures: a decrease in the order in the alloy begins already at $T \approx 300$ K in the case of TAPBs in the <100> direction and at ~1100 K in the case of SAPBs in the <110> direction. In the temperature range from ~1100 to ~1800 K there is a stable decrease in the value of the long-range order parameter. With a further increase in temperature, the long-range order tends to zero, which indicates a rapid disordering in the system. The temperature range of long-range order variations in the system (Fig. 2.23 b) is consistent with the range of configuration energy changes (Fig. 2.21).

Of great interest are the changes in the atomic and domain structure of the intermetallic compound – monoaluminide NiAl with the APB complex with increasing temperature of the system. First, we consider the changes in the atomic and domain structure of an intermetallic compound with a complex of thermal APBs in the <100> direction as a function of temperature in the region of low-stability pre-transitional states with increasing temperature. Then we will carry out exactly the same consideration of the changes in the atomic and domain structure of a system with a complex of shear antiphase boundaries in the <110> direction as a functio of temperature. Next, we analyze the influence of the type of antiphase boundaries and the plane of occurrence of antiphase on the structure-phase features of the system as a function of temperature in the region of low-stability pre-transitional states.

Let us consider the change in the atomic and domain structure of an intermetallic compound with a complex of thermal APBs in the <100>

direction as a function of temperature in the region of low-stability pre-transitional states, which is shown in Figs. 2.24 and 2.25, respectively.

It is easy to see that even at low temperatures (~300 K), significant structural order disturbances occur at the Al–Al interface, which increase with increasing temperature.

At the Ni–Ni interface, the first disordered regions appear at a higher temperature (~600 K). This can be understood as a significant role of the size factor (the atomic size of Al is much larger than the corresponding size of Ni (Table 2.1)) in the formation of the structural-phase low-stability state of the system. In the temperature range 600–1400 K, faceting and blurring of the boundaries are observed. At ~1000 K, disordered regions begin to appear throughout the system; the shape and size of the boundaries change. The disordered regions continue to increase throughout the system with a further increase in temperature, and the antiphase boundaries disappear ($T \approx 1600$ K). At higher temperatures, almost the entire alloy is disordered, only small-sized domains remain. It should be noted that the range of changes in the atomic and domain structure of the system and low-stability states corresponds to the range of changes in the configurational energy of the alloy with TAPBs (Fig. 2.21) and the contribution of APBs to the energy of the alloy (Fig. 2.22).

Let us consider the change in the atomic and domain structure of a system with a complex of shear antiphase atoms in the <110> direction as a function of temperature in the region of low-stability pre-transitional states, which is shown in Figs. 2.26 and 2.27, respectively.

Up to 600 K there are no structural disturbances. The first insignificant changes occur in the regions of the boundary intersection crossing in the temperature range from 600 to 1000 K. With an increase in temperature (~1100 K), the first disordered regions appear throughout the system, but there are no changes in the region of antiphase boundaries occur, the boundaries remain stable. In the range 1200–1400 K, the number and size of disordered regions throughout the system increases. In the range of 1400–1500 K, faceting and blurring of the boundaries are observed; the number and size of disordered regions increases with temperature. The regions of boundary intersection tend to transition to an ordered state. A packing defect (a monoatomic column of one of the alloy components of the size of one unit cell) is formed in the region of the border crossing, which does not correspond to a simple shear boundary, therefore, its behaviour differs from the that of the SAPBs themselves. These differences provoke an initial rearrangement into an ordered state, while the SAPBs themselves remain stable. At $T \approx 1600$ K, the influence of APBs is still noticeable, but it becomes less pronounced.

Naturally, at temperatures above the temperature of a hypothetical order–disorder phase transition, the antiphase boundaries disappear, the system is disordered, and only small-sized domains are observed. The situation is similar to that observed in a defect-free alloy and an alloy with a complex of thermal APBs in the <100> direction. It should be noted that the range of changes in the atomic and domain structure of the intermetallic compound corresponds to the range of changes in the configurational energy of the alloy (Fig. 2.21) and the contribution of the APH to the energy of the system (Fig. 2.22).

Conclusion. An analysis of the effect of complexes of antiphase boundaries in the <100> (a pair of thermal APBs) and <110> (a pair of shear APBs) directions on low-stability structural-phase states of an intermetallic compound (using the example of an intermetallic monoaluminide NiAl) has shown that the presence of antiphase boundaries has a significant effect on the structural-phase state and energy characteristics of the entire system.

The most significant for the long-range order in the system is the very appearance of a defect in the form of APBs; the difference in the type and plane of their occurrence does not significantly affect the behaviour of the long-range order with temperature. Naturally, a system with structural defects is less ordered than a defect-free system, which is manifested in the behaviour of the $\eta = f(T)$ curves: the curve of the defect-free intermetallide lies higher than the curves of an alloy with APBs. The presence of a defect in the form of APBs contributes to the onset of disordering of the system at lower temperatures: a decrease in order begins in the case of thermal APBs in the <100> direction at a lower temperature compared with the case of shear APBs in the <110> direction.

In the NiAl intermetallic compound with a complex of thermal APBs in the <100> direction, the first structural order disturbances always appear near the Al–Al interface. In an alloy with a complex of shear APBs in the <110> direction, structural order disturbances at low temperatures are observed only in the regions of boundary intersection. The presence of APBs affects the stability of the system during heating.

At moderate temperatures, the NiAl monoaluminide without structural defects is more stable than an alloy with APBs. It is shown that the disordering process is accompanied by blurring of the boundaries and their faceting.

The effect of APB complexes on the structural energy characteristics of the entire system during heating is significant in the region of low-stability structure-phase states of the intermetallic compound to the structural phase transformation temperature, which is consistent with previous studies [5–18, 21, 22].

Summary

Using the Monte Carlo method it has been shown on the example of the NiAl intermetallic compound of the Ni–Al system that irreversibility of processes is observed during thermal cycling in the course of structure-phase transformations in BCC intermetallic compounds. As a result of the heating and cooling cycle, there is a kind of hysteresis, the presence of which indicates the irreversibility of the processes which implies a difference in the structure-phase states at the heating and cooling stages. Analysis of the atomic and phase structure of the system in the course of heating and cooling, i.e. in the process of hypothetical order–disorder and disorder–order phase transitions, confirmed the difference in structure-phase states at the stages of heating and cooling. It is shown that for the order–disorder transition to occur, the system needs to be slightly overheated relative to the traditionally understood phase transformation temperature, and for the disorder–order transition to occur, the system needs to be slightly supercooled relative to the same temperature.

This is consistent with the temperature dependences of energy and order. After the completion of the disorder–order phase transition in the system, two antiphase domains of the $B2$ superstructure are formed.

Conclusion. Using the Monte Carlo method, it was shown that irreversibility of processes is observed during thermal cycling during structural-phase transformations in bcc alloys, for example, NiAl intermetallic compounds of the Ni–Al system. As a result of the heating and cooling cycle, a kind of hysteresis is realized, the presence of which indicates the irreversibility of the processes that take place, which implies a difference in the structure-phase states at the heating and cooling stages. Analysis of the atomic and phase structure of the system in the processes of heating and cooling, i.e. during the phase transition, order – disorder and disorder – order, confirmed the difference in structure-phase states in the heating and cooling stages.

Thus, for the order–disorder transition to occur, the system needs to be slightly overheated with respect to the traditionally understood phase transformation temperature, and for the order–disorder transition the system needs to be slightly supercooled relative to the same temperature.

References

1. Kositsyn S.V., Kositsyna I.I. // Advances in metal physics. - 2008. - V.9. - P. 195–258.
2. Potekaev A.I., Starostenkov M.D., Kulagina V.V. The effect of point and planar defects on structural-phase transformations in the pre-transition weakly stable region of metal systems / under the total. ed. A.I. Potekaev. - Tomsk: NTL Publishing House, 2014 .- 488 p.
3. Koneva N.A., Trishkina L.I., Potekaev A.I., Kozlov E.V. Structural-phase transformations in weakly stable states of metallic systems during thermosilic interaction / under the general. ed. A.I. Potekaeva. - Tomsk: NTL Publishing House, 2015 .- 344 p.
4. Chaplygin P. A., Potekaev A. I., Chaplygin A. A. et al. // Izv. Univ.. Fizika. - 2015. - V. 58. - No. 4. - P. 52–57.
5. Chaplygina A.A., Chaplygin P.A., Starostenkov M.D. et al. // Fundamental problems of modern materials science. - 2016. - V. 13. - No. 3. - P. 403–407.
6. Potekaev A. I., Chaplygin A. A., Chaplygin P. A. et al. // Izv. Univ. Fizika. - 2016. - V. 59. - No. 5. - P. 3–8.
7. Potekaev A.I., Chaplygina A.A., Starostenkov M.D. et al. // Izv. Univ. Fizika. - 2012. - V. 55. - No. 7. - P. 78–87.
8. Potekaev A.I., Klopotov A.A., Kozlov E.V., Kulagina V.V. // Izv. Univ. Fizika. - 2011. - V. 54. - No. 9. - P. 59–69.
9. Potekaev A.I., Chaplygina A.A., Kulagina V.V. et al. // Izv. Univ. Fizika. - 2017. - V. 60. - No. 2. - P. 16–26.
10. Potekaev A.I., Klopotov A.A., Trishkina L.I. et al. // Bulletin of the Russian Academy of Sciences. Ser. Physical Physics. - 2016. - V. 80. - No. 11. - P. 1576–1578.
11. Chaplygina A.A., Potekaev A.I., Chaplygin P.A. et al. // Fundamental problems of modern materials science. - 2016. - V. 13. - No. 2. - P. 155–161.
12. Chaplygina A.A., Potekaev A.I., Chaplygin P.A. et al. // Izv. Univ. Fizika. - 2016. - V. 59. - No. 10. - P. 13–22.
13. Poletaev G.M., Potekaev A.I., Starostenkov M.D. et al. // Izv. Univ. Fizika. - 2015. - V. 58. - No. 1. - P. 38–43.
14. Potekaev A.I., Morozov M.M., Klopotov A.A. et al. // Izv. Univ.. Ferr. Metallurgiya. - 2015. - V. 58. - No. 8. - P. 589–596.
15. Potekaev A.I., Chaplygina A.A., Chaplygin P.A. et al. // Izv. Univ. Fizika. - 2017. - V. 60. - No. 9. - P. 118–126.
16. Potekaev A.I., Chaplygina A.A., Chaplygin P.A. et al. // Izv. Univ. Fizika. - 2017. - V. 60. - No. 10. - P. 115–124.
17. 17. Potekaev A.I., Chaplygin A.A., Chaplygin P.A. et al. // Izv. Univ. Fizika. - 2018. - V. 61. - No. 3. - P. 12–27.
18. Potekaev A.I., Chaplygina A.A., Chaplygin P.A. et al. // Izv. Univ. Fizika. - 2019.- V. 62. - No. 1. - P. 104–111.
19. Iveronova V.I., Katsnelson A.A. Short-range order in solid solutions. - M .: Nauka, 1977 .-- 253 p.
20. Krivoglaz M.A., Smirnov A.A. Theories of ordered alloys. - M .: Fizmatgiz, 1958.- 388 p.
21. Potekaev A.I., Chaplygina A.A., Chaplygin P.A. et al. // Izv. Univ. Fizika. - 2019.- V. 62. - No. 2. - P. 123–132.
22. Potekaev A.I., Chaplygina A.A., Chaplygin P.A. et al. // Izv. Univ. Fizika. - 2019.- V 62. - No. 3. - P. 117–124.

3

Pre-transitional low-stability states in TiNi-based alloys

Using the influence of point defects and their complexes on structural transformations in TiBi-based alloys, it is shown that in the low-stability state of a condensed system (in TiNi, these are the so-called pre-transitional states), the interaction of structural defects can have a significant effect on the structural-phase transformations, and nanoscale objects play a very important role in phase stability in the field of martensitic transformations (MTs).

It is found that the concentration dependences of the onset temperature M_s of the direct MT in stressed and unstressed specimens have different values and different functional dependences. This allows us to hypothesize that the observed effect is associated with the presence of low-stability pre-transitional states in the region of the MT in alloys of this class, and the structure and properties of these states depend on the previous thermomechanical treatments of the system.

The results of studies of the physical properties in multicomponent Ti (Ni, Co, Mo)-alloys, the effect of annealing and thermal cycling on the MT intervals allowed us to establish that thermal cycling through the MI region in microalloyed alloys leads to a slight decrease in the temperature of the onset of the MT and a noticeable increase in the area under the temperature curve of electrical resistivity with saturation after the 20th cycle.

TiNi-based alloys are most successful for creating materials with shape memory effects (SME) [1]. The choice of alloys to design new functional materials with SME is determined by a set of optimal properties. These, in particular, are the MT temperature ranges, the degree of SME, the value of the martensitic shear stress, strain and thermal deformation, etc. [1–4]. In order to form some optimal target physical and mechanical properties of materials, it is advisable to study the alloys based on TiNi, which is the most well-known SME material. Structures made of TiNi-based alloys are often exposed to external stress and temperature during their service life. The greatest challenge in designing such structures is to ensure the stability and predictability of the material properties. It is known that the critical temperatures and critical stresses for the $B2 \leftrightarrow B19'$ martensitic transformation are very sensitive to both thermal and mechanical cycling [5]. During thermal cycling through the MT region, defects accumulate in alloys, which is called 'phase hardening' [1]. These structural defects play a very significant role in the field of structural phase transformations. A study of variations in the parameters characterizing the physicomechanical properties of alloys in the MT region is a necessary aspect of understanding the nature of the influence of thermal cycling on the properties of alloys.

It has long been established [6] that external thermal force impacts have a significant effect on thermoelastic martensitic transformations in alloys. This effect extends to different stages of transformation: pre-transitional low-stability states of the system, stages of nucleation and growth of martensite crystals. It should be specially noted that TiNi-based alloys are characterized by the presence of a hysteresis loop during thermal cycling through the MT region, and the loop changes with varying number of cycles [1, 6–9]. For example, the temperature dependences of the electrical resistivity curves through the region of martensitic transformations in TiNi-based alloys exhibit pronounced asymmetry in the direct and reverse transformations. It is assumed [1, 6–9] that this behaviour of electrical resistivity in the region of martensitic transition is due to the different nature of the temperature dependences of electrical resistivity in the austenitic and martensitic phases, which allows us to explain this pronounced asymmetry. A validated assumption was made [8, 9] that the presence of a hysteresis indicates the realization of a number of low-stability pre-transitional structural states in the transformation region. However, on the other hand, the presence of low-stability structural states does not necessarily lead to the appearance of a hysteresis loop. The situation in the pre-transitional low-stability region is somewhat similar to the situation of long-period structures near the phase transformation temperature [10–14].

Thus, TiNi-based alloys are very important in understanding the nature of the effect of various influences on the structure and properties of materials, since they have extraordinary functional properties, primarily, shape memory effects and superelasticity. Materials with these functional properties are widely used [5]. SME in TiNi-based alloys is due to thermoelastic martensitic transformations occurring in them. An important stage was the understanding that the anomalies in the behaviour of the structural and physicomechanical properties in alloys with thermoelastic MTs (under thermal stresses) are caused by pre-transitional (pre-martensitic) low-stability states of the crystal lattice before the MTs [15, 16]. The study of the features of the resulting anomalies allows us to reveal the physical nature and microscopic mechanisms of the preparation of the crystal structure for the upcoming MT.

3.1. The effect of point defects on structural-phase transformations in low-stability states of functional materials

It is known that in the vicinity of structural-phase transformations the concentration and role of the structural point defects is large, therefore, it is necessary to study the effect of these defects and their complexes on the nucleation of the martensitic phase in the pre-martensitic region (in the low-stability state), i.e. before the MT of the Ti–Ni system in the region of equiatomic composition.

From the point of view of the influence of structural defects on the MTs, the latter can be conditionally divided into two groups: transformations of the first kind and those close to the second kind [15–20]. Of interest are, first of all, alloys belonging to the second group. These primarily include the alloys based on precious metals, TiNi-based and other alloys. The martensitic phase in these alloys nucleates on complex structural defects such as grain boundaries, interphase boundaries, free surfaces, complexes of point defects, etc. Similar defects, being local stress concentrators, lower the activation barrier of the martensitic phase nucleation.

In the alloys of this group, pre-transitional anomalies are most pronounced. This is manifested in the anomalous behaviour of the electrical resistivity, the coefficient of internal friction, the shear moduli, the appearance of dips on the phonon dispersion curves at characteristic values of the wave vector; diffuse scattering and extra reflections are observed in X-ray and electron diffraction patterns, and the electronic spectrum is restructured [21, 22].

It was shown [15, 16] that simple defects in the crystal structure (dislocations, stacking faults, and complexes of point defects) and their interaction with soft phonon modes can play an important role in the nucleation of the martensitic phase. Defects can be the nucleation or pinning centres for the regions with short-range order of atomic displacements and intermediate shear structures.

Order parameter, defects and *B*2 crystalline structure. The lattice of the *B*2 structure is a simple cubic (point group O_h, space group O_h^1). In the *B*2 structure (Fig. 3.1 *a*) one of the sublattices ($i = 1$), composed of α-type nodes with coordinate (000), is occupied by atoms of sort *A*; the second sublattice ($i = 2$) is made up of nodes of β-type nodes with a coordinate $\left(\frac{1}{2}, \frac{1}{2}, \frac{1}{2}\right)$ that is replaced by atoms of variety *B*. In the case when atoms *A* or *B* are on 'foreign' sublattices, such defects are called anti-structural. The second group of lattice defects is formed when only *B* atoms occupy the sites

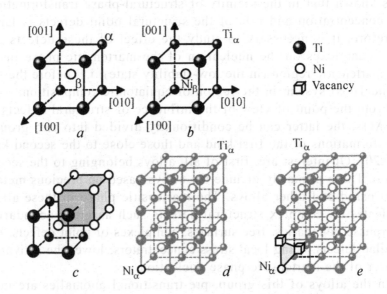

Fig. 3.1. Unit cell of the *B*2 structure phase with selected α- and β-nodes (*a*). In the TiNi phase with the *B*2 structure (*b*): Ti atoms are located in the α-nodes, and Ni atoms are located in the β-nodes. Two unit cells of a simple cubic lattice are inserted into each other (*c*). Computational grid of 12 unit cells of *B*2 structure with anti-structural defect Ni_α (*d*) and a triple defect (*e*).

of the sublattice α. These are termed the triple defects. For the stoichiometric composition, there are simultaneously three types of defects: two vacancies in the $B2$ sublattice and one B atom in the α sublattice site (Fig. 3.1 e).

The crystal lattice of an ordered solid solution is represented in the form of several sublattices, each of which is occupied by atoms of a certain kind [23]. The $B2$ structure can be represented in the form of two simple cubic lattices embedded in each other, in which atoms of different kinds A and B can purely geometrically occupy equally equivalent positions (Fig. 3.1 c). To describe the distribution of atoms over the nodes of the crystal lattice of an ordered alloy, we use the long-range order parameter (η) according to the Gorsky–Bragg–Williams approximation [23]. The degree of atomic long-range order in ordered alloys is described using the probability of substitution of nodes of different types belonging to different sublattices:

$$\eta = \frac{P_i^A - C_A}{1 - v},$$

where C_A is the concentration of atoms of variety A; P_i^A is the probability of finding an A atom in a node i; v is the fraction (concentration) of nodes of the i-th type. In a completely disordered alloy, the nodes of the crystal lattice are indistinguishable, the probability of detecting an A atom in any of them is equal to the concentration C_A and, therefore, $\eta = 0$. In a completely ordered alloy of stoichiometric composition, $C_A = v$ and $P_i^A = 1$, then $\eta = 1$. With partial ordering, $0 < \eta < 1$.

For the TiNi phase of stoichiometric composition with the $B2$ superstructure, we have $C_A = 0.5$, $v = 0.5$, and $P_i^A = 1$. As the alloy becomes disordered, the value of P_i^A decreases, the concentration of anti-structural defects increases and becomes equal to $1 - P_i^A$. The probability of their formation depends on the excess energy (ΔE) of their formation:

$$\Delta E^{\text{Ni}} = E_\beta^{\text{Ni}} - E_\alpha^{\text{Ni}} \text{ and } \Delta E^{\text{Ti}} = E_\beta^{\text{Ti}} - E_\alpha^{\text{Ti}}.$$

When an atom occupies a 'foreign' node, its neighbourhood changes, and so does the number of 'right' and 'wrong' pair bonds. In Fig. 3.1 a, b in the $B2$ structure, all atoms in the $B2$ structure are in the right positions. The coordination number on the first coordination sphere is $Z_1 = 8$. For Ni atoms located in the β-nodes, all eight neighbors are Ti atoms, and for Ti atoms located in the α-nodes, there are also eight Ni atoms in the first coordination sphere in the case of complete ordering. In the formation of anti-structural defects, Ni–Ti pair bonds are partially replaced by Ni–Ni, i.e. 'wrong' bonds appear.

In Fig. 3.1 *d*, the arrangement of atoms is almost the same as in Fig. 3.1 *b*, except that the Ni atom is located in the α-node with the (100) coordinate. The formation of an anti-structural defect should lead to a decrease in the degree of order and, as a consequence, to a change in the temperature of the MT onset.

The effect of point defects on martensitic transformations is evidenced by the significant dependence of the characteristics of martensitic transformations on the concentration of the alloying element and thermomechanical treatment.

The different roles of the two sublattices in the formation of vacancies is manifested in the kinetics and morphology of the formation of particles with Ti_2Ni and $TiNi_3$ phases in the TiNi matrix phase and, as a result, determines the main characteristics of martensitic transformations and the physico-mechanical properties of TiNi-based alloys. Another important factor is not only the presence, but also the distribution of defects in the high-temperature matrix. It is known that in Ti–Ni alloys enriched with nickel, depending on the heat treatment, the sequence of transformations changes from $B2 \rightarrow B19'$ to $B2 \rightarrow R \rightarrow B19'$ in the annealed alloys. It is assumed that this fact is associated with the appearance of fields of rhombohedral elastic stresses during the precipitation of Ni_4Ti_3 particles or with the ordering of excess nickel atoms in the sites of the titanium sublattice in annealed alloys [24].

Results and its discussion. The TiNi-based alloys have a large number of structural (primarily, point) defects associated with a deviation from the stoichiometric composition or with temperature-induced disordering. The presence of structural defects is clearly manifested in TiNi binary alloys in the region of equiatomic composition on the behaviour of structural and physical parameters depending on concentration. It has been found out that in a TiNi stoichiometric alloy the electrical resistivity has a minimum value, and the density monotonically depends on the concentration (Fig. 3.2). The non-monotonic nature of the change in structural and physical parameters in the region of equiatomic composition

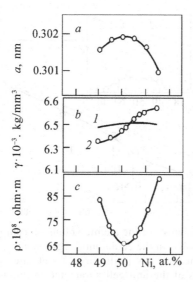

Fig. 3.2. Concentration dependences: a – lattice parameter; b – density (1 – theoretical curve, 2 – experimental); c – electrical resistivity.

undoubtedly reflects not only the presence of point defects in the crystal lattice, but also the different contribution of vacancies and anti-structural defects depending on the direction of deviation from the stoichiometric composition [15].

It is worth noting the deviation of the experimental values of the atomic volumes of the alloys from the Zen law in the case where the composition of the TiNi alloy deviates from the equiatomic composition (Fig. 3.3). There is a significant discrepancy between the theoretical and experimental values of atomic volumes associated with superstructural compression, which occurs during the formation of intermetallic compounds with high long-range order values [16]. A nonlinear change in the atomic volume in the region of equiatomic composition is due to a deviation from the stoichiometric composition.

Comparing these data with the results of measurements of electrical resistivity, density, lattice parameters and atomic volume (Figs. 3.2 and 3.3), we can conclude that the concentration of vacancies in TiNi alloys is significant when their compositions deviate from stoichiometry.

Role of interatomic interaction potentials [19, 20, 25–27]. The interatomic interaction potential reflects the energy of this interaction; therefore, the interatomic interaction potentials between

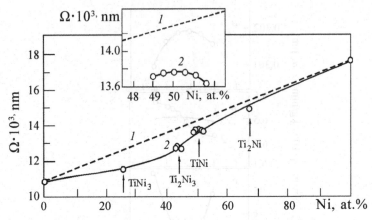

Fig. 3.3. Concentration dependences of the atomic volume in compounds of the Ti – Ni system (inset – $\Omega = f$(at.% Ni) in the equiatomic region): curve 1 – theoretically calculated Ω according to the Zen law [16]; curve 2 - atomic volume calculated from the experimental values of the unit cell parameters of the compounds in the TiNi system.

the like (Ni–Ni and Ti–Ti) and unlike (Ni–Ti) atoms make it possible to estimate the configurational energy of the crystal system, i.e. estimate the arrangement of atoms along the nodes of the crystal lattice.

According to the data on the wave functions of Ti and Ni atoms, it follows from calculations [25] that the interaction of titanium atoms in the TiNi compound occurs over a longer range than the interaction of nickel atoms. This is consistent with the data that the wave functions of Ti atoms are more diffuse than those of Ni atoms. The calculation of the interatomic interaction potentials in the TiNi compound ($\varphi_{NiNi}(r) > \varphi_{NiTi}(r) > \varphi_{TiTi}(r)$), made in the framework of the pseudopotential theory [25], showed that the interaction potential $\varphi_{TiTi}(r)$ has a deep minimum between the first and second coordination areas (Fig. 3.4). A shallow minimum in this region is observed on the pair potential curve $\varphi_{NiNi}(r)$. Such behaviour of the potentials indicates that the energy of vacancy formation at the sites of the nickel atom sublattice is 0.16 eV lower than that at the sites of the titanium atom sublattice, which leads to a higher concentration of vacancies on the nickel sublattice. It follows from the shape of the curves of pairwise interaction potentials that the energy of the nickel atom located in the site of its sublattice is higher than the energy of this atom placed in the site of the 'titanium' sublattice. It is energetically more profitable for a nickel atom to be in the site of the titanium sublattice, while at the same time it is beneficial for

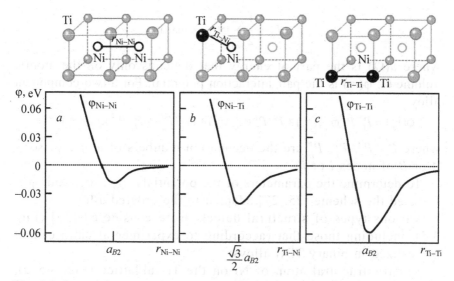

Fig. 3.4. Potential curves of interatomic interaction: a – between the like atoms Ni–Ni; b – between the unlike atoms of Ni–Ti; c - between Ti–Ti.

the titanium atom to be in the site of its sublattice. From here one can expect that vacancies should be formed mainly on the 'nickel' sublattice in the TiNi compound.

The experimental data obtained from the measurements of electrical resistivity, density, lattice parameters and theoretical estimates allow us to conclude that vacancies are predominantly located on the nickel sublattice in the $B2$ structure. Therefore, with an increase in the concentration of titanium, the concentration of vacancies in the –nickel' sublattice increases, and with an increase in the concentration of nickel, the concentration of nickel atoms in the titanium sublattice increases.

Effect of point defects and their complexes on the structure-phase state of a crystalline system [19, 20, 25–27]. Despite the fact that a number of models for the nucleation of the martensitic phase (for example, the Yamada model [28]) have been proposed to explain many aspects of the microscopic mechanism of the magnetic field, none of them can describe in detail the paths of the transformation reaction and the mechanism of inheritance of high-temperature sdtructure defects by the martensitic phase.

The authors of [26] attempted to solve these problems using the method of computer simulation. To this end they presented the energy of the system as the sum of the pair and volume components:

$$U = \frac{1}{2} \sum_{i \neq j} \varphi(r_{ij}) + u(\Omega),$$

where $u(\Omega)$ is the part of energy that depends only on the atomic volume Ω; $\varphi(r_{ij})$ is the pair interaction potential. For a two-component alloy

$$\varphi(r_{ij}) = P_i^A P_j^A \varphi_{AA}(r_{ij}) + P_i^B P_j^B \varphi_{BB}(r_{ij}) + (P_i^A P_j^B + P_i^B P_j^A)\varphi_{AB}(r_{ij}),$$

where P_i^A, P_j^A, P_i^B, P_j^B are the occupation numbers of atoms of sorts A or B of nodes i or j.

To determine the parameters of the potentials φ_{AA}, φ_{BB} and φ_{AB}, we used the scheme [25, 27] adapted to the ordered alloy.

Various types of structural defects were considered [26] (Fig. 3.5), including those that (according to experimental data) cannot be realized in binary TiNi alloys:

1) anti-structural atom of Ni on the Ti sublattice (Fig. 3.5 *a*); 2) a defect, caused by disordering, by the rearrangement of Ti and Ni atoms on the first coordination sphere (Fig. 3.5 *b*); 3) an anti-structural Ni atom and an order defect forming a linear chain along the <111> direction (Fig. 3.5 *c*); 4) an excess Ni atom and an order defect located in the <110> plane (Fig. 3.5 *d*).

It has been found out that in the course of ordering of the anti-structural nickel atoms in the sites of the titanium sublattice, the *B*2 structure remains stable. In cases 2–4, the displacement fields near the defects lose cubic symmetry; their interaction with each other contributes to the instability of the *B*2 lattice and to the formation of hexagonal martensite through a chain of transformations. With

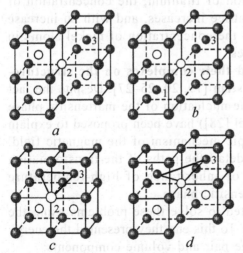

Fig. 3.5. Different types of structural defects.

the indicated types of defects, the same final structure is obtained if the fine structure of the defects is not taken into account. It is a hexagonal layered ordered phase of alternating planes containing atoms of the same kind. The layers are perpendicular to the c axis, oriented along the <111> direction of the $B2$ lattice, and the axes a and b are parallel to $<\overline{1}\,\overline{1}2>$ and $<\overline{1}2\overline{1}>$.

Figure 3.6 shows that the {111} BCC lattice planes are displaced in the <111> direction similar to how this occurs when the ω phase is formed. As the system reached equilibrium, a local minimum was recorded, which corresponds to the formation of an intermediate structure. The unit cell of the intermediate structure contains 18 atoms. The c axis is directed along <111> and is equal to $a\sqrt{3}$ (a is the lattice constant), the axes a and b are equal to $a\sqrt{6}$ and are oriented along the $<\overline{1}\,\overline{1}2>$ and $<\overline{1}2\overline{1}>$ directions, respectively.

The transformation into a hexagonal structure, in addition to the type of defect introduced, is significantly affected by boundary conditions. A transition to a phase with hexagonal symmetry becomes possible if the computational cell is a multiple of $3n \times 3n \times 3n$, where n is an integer. In this case, in the <111> direction an integer number of modulating displacement waves is stacked and the boundary conditions do not impede the transfer of the martensitic shift from one cell to another. If these conditions are not met, a structural transition was not observed.

Thus, in the pre-transitional low-stability state of the TiNi compounds with low elastic moduli, long-range displacement fields develop around point defects, therefore, even at their low concentration, the defects interact with each other. In the case of a pre-transitional low-stability state, they cannot be considered as isolated defects. The forming static displacement fields can both stabilize the $B2$ structure and contribute to its instability and martensitic phase transition. In the considered model, in the presence of a certain type of defects in the $B2$ structure, the latter became unstable to displacements of the {111} planes along the <111> type direction.

The transformation of a material with point defects and their complexes leads to the disruption of the order in the arrangement of atoms; therefore, the martensitic phase stores additional chemical energy. Consequently, energy causes appear in the lattice for the reverse martensitic transformation with a certain orientation of the transformation path. In the 'exactly backward' reaction, a complete

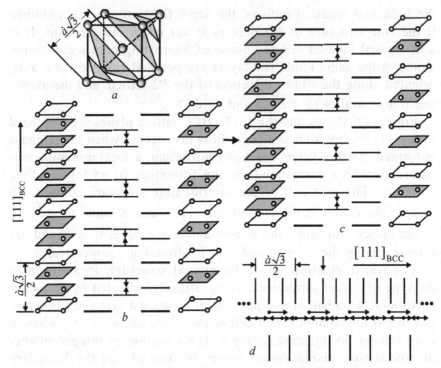

Fig. 3.6. Unit cell of a BCC lattice with distinguished planes of type (111) (*a*). The displacement scheme of the {111} atomic series upon the formation of a ω-like structure without (*b*) and with a defect (*c*) in the lattice. The horizontal lines show the positions of the {111} planes before (the left side of the circuit) and after the A1–ω transition (right). A simplified diagram of the displacements of the {111} atomic rows upon the formation of a ω-like structure .(d): vertical lines — the positions of the {111} planes before (above) and after (below) the transition.

restoration of the initial structure with low energy occurs. A change in the number of defects during the martensitic transition leads to the accumulation by the defects of the energy associated with the chemical contribution to the thermodynamic potential, since the number of pairs of atoms of the same type in the first coordination spheres changes.

The results of a model experiment show that during martensitic transformation from the *B*2 phase into a ω-like structure, high-energy linear chains of point defects in the $<111>_{B2}$ direction are formed, which in some cases leads to an increase in their number in the martensitic phase. This fact can play a significant role in determining the reverse path of the martensitic transformation.

Conclusion. In the pre-transitional low-stability state of systems (in the case under consideration, states with low elastic moduli), significant static displacements of atoms from the sites of the crystal lattice take place before the transformation. As a result of this, even at a low concentration of defects, they interact with each other.

In the final structure, atomic displacements in the vicinity of the defect are localized. The fields of atomic displacements in the vicinity of defects not lying in the {110} type planes prevent shuffling of the planes in the {1$\bar{1}$0} direction and the implementation of the BCC–FCC martensitic transformation by this mechanism. In this case, a different variant of the development of martensitic transformation is not ruled out. Thus, the interaction of defects located in the <111> planes leads to the appearance of a long-period ω-like structure, which is formed by shuffling displacements of the <111> atomic rows. Thus, in the pre-martensitic state, when the system is at the threshold of its stability, the interaction of the atomic displacement fields developing in the vicinity of defects can affect the choice of a possible martensitic transition path.

Point defects caused by a deviation from stoichiometry and a disruption of the long-range order in the arrangement of atoms can affect the stability of the $B2$ structure and contribute to the transformation of the martensitic type into a ω-like structure. In this case, high-energy chains of point defects are formed in the <111> direction, which leads to an increase in the number of point defects in the martensitic phase. The latter can have a significant role in determining the return path of the martensitic transition.

Thus, by studying the effect of point defects on structural transformations in TiNi-based alloys, it has been shown that in the low-stability state of a condensed system, the interaction of structural defects can have a significant effect on the structural-phase transformations, and nanoscale objects play a very important role in the stability of the crystal lattice in the region of MTs.

3.2. The effect of deformation on the temperature regions of martensitic transformations in TiNi-based alloys

External thermal force impacts have a significant effect on thermoelastic MTs in alloys [6]. The effect is carried out in different stages of transformation: pre-transitional low-stability states of the system and the nucleation and growth of martensite. For TiNi-based alloys, thermal cycling through the region of martensitic transformations is characterized by the presence of a hysteresis loop, which undergoes changes with varying number of cycles [1, 6–9]. Thus, in the TiNi-based alloys, upon thermal cycling through the MT region, the temperature curve of the electrical resistivity exhibits pronounced asymmetry in the direct and reverse transformations. It is reasonably assumed [1, 6–9] that this behaviour of electrical resistivity in the vicinity of the martensitic transition is due to the different nature of the temperature dependences of electrical resistivity in the austenitic and martensitic phases, which allows us to explain this pronounced asymmetry. In fact, a validated assumption was made [8, 9] that the presence of hysteresis indicates the realization of a number of low-stability (pre-transitional) structural states in the transformation region.

In metals and alloys in which MT occurs, plastic deformation and external applied stresses at temperatures above M_s can themselves initiate transformations. Drawing an analogy with the behaviour of low-stability states in long-period ordered structures, we can assume that the thermomechanical action would lead to an increase in the temperature of structure-phase transformations. To this end, we study the effect of deformation on the temperature ranges of martensitic transformations in alloys based on the equiatomic TiNi alloy.

Materials and experimental procedures. In this study we investigate TiNi-based alloys (e.g., TN-10), doped with 99.99% pure niobium. The alloys were melted in an argon atmosphere in an induction furnace. The weight loss during smelting did not exceed 0.01%. The produced ingots were cut by electrospark cutting to form specimens measuring $1 \times 1 \times 120$ mm which were then annealed at 900°C for 0.5 h for homogenization, followed by quenching in ice water. The electrical resistivity was measured by the standard four-point method.

The alloys of a given composition were examined by the method of X-ray diffraction analysis and measurement of the temperature dependence of the electrical resistivity to determine the characteristic temperatures of the magnetic field from the position of the martensitic points, which depend on the concentration.

Under the condiyions of external thermal loading, a cooling-heating cycle was carried out in the temperature range of the MT on the laboratory setup. As a result of the experiment, the temperature dependences of the strain were obtained.

Results and discussion. To study the physical properties of Ti (Ni, Nb, Mo) alloys doped with different concentrations of Nb the

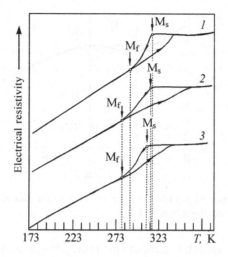

Fig. 3.7. Temperature dependence of electrical resistance ρ on temperature in alloys $Ti_{50}Ni_{49.2}Mo_{0.3}Nb_{0.5}$ (curve 1), $Ti_{50}Ni_{48.7}Mo_{0.3}Nb_1$ (curve 2) and $Ti_{50}Ni_{48.2}Mo_{0.3}Nb_{1.5}$ (curve 3).

temperature dependences of the electrical resistivity $\rho = \rho(T)$ were experimentally obtained, which are shown in Fig. 3.7. From these curves and X-ray analysis data it follows that in these alloys with decreasing temperature a martensitic transition occurs from the initial $B2$ phase to the martensitic phase.

The above dependences $\rho = \rho(T)$ show that, with an increase in the concentration of niobium, the MT shifts to lower temperatures and the shape of the electrical resistivity curves changes. From the nature of the change in the dependences $\rho = \rho(T)$, the characteristic

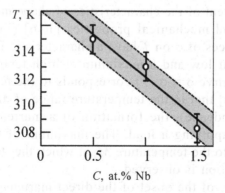

Fig. 3.8. Dependence of temperature M_s on Nb concentration in Ti(Ni, Nb, Mo) alloys, obtained from temperature curves of electrical resistivity. The dashed lines show the region where M_s tends to decrease as a function of Nb concentration..

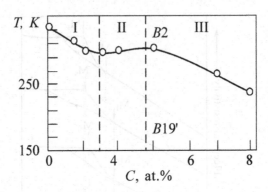

Fig. 3.9. Dependence of the temperature M_s on the niobium concentration in Ti (Ni, Nb) alloys quenched from 800°C [29].

temperatures of the direct martensitic transformation were determined, which made it possible to construct the dependence of the temperature M_s on the concentration of niobium in the alloy (Fig. 3.8). It is easy to see that doping with niobium of an alloy, for example TN-10, leads to a noticeable shift of the temperature range of martensitic transformations to lower valuesw. Moreover, the results obtained on the four-component alloy Ti(Ni, Nb, Mo) correlate well with the literature data on the three-component alloy Ti(Ni, Nb) [35] (Fig. 3.9).

To study the mechanical properties of $Ti_{50}Ni_{49.7-x}Mo_{0.3}Nb_x$ alloys ($x = 0.5$ at.%; 1 at.%; 1.5 at.%), the temperature dependences of critical martensitic shear stresses and yield strength were experimentally obtained. The study of temperature dependences of martensitic shear stresses in Ti(Ni, Nb, Mo) alloys allows a qualitative assessment of the characteristics of shape memory effects, superelasticity, and mechanical properties [1, 6] (Fig. 3.10). The obtained dependences of σ on T have a characteristic form of curves with a minimum at low and a maximum at high temperatures. The minimum on the curve $\sigma = \sigma (T)$ corresponds to the temperature M_s. It is believed [1, 6] that in the temperature range of M_s (a minimum on the $\sigma(T)$ dependence), the formation of a martensitic phase is possible without applying a load. The maximum of the curve $\sigma = \sigma (T)$ corresponds to the temperature M_d at which the 'true' resistivity to plastic deformation is observed.

The temperature of the onset of the direct martensitic transformation M_s, obtained from the data $\sigma = \sigma(T)$, has different values and a different functional dependence in comparison with the values observed in unstressed specimens and the $\rho = \rho(T)$ dependences

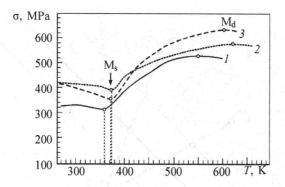

Fig. 3.10. Temperature dependences of martensitic shear stresses in Ti(Ni, Nb, Mo) alloys with different Nb concentrations: curve 1 – $Ti_{50}Ni_{49.2}Mo_{0.3}Nb_{0.5}$; curve 2 – $Ti_{50}Ni_{48.7}Mo_{0.3}Nb_{1}$; curve 3 – $Ti_{50}Ni_{48.2}Mo_{0.3}Nb_{1.5}$.

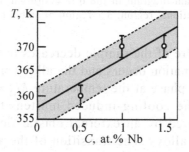

Fig. 3.11. Concentration dependence of the temperature of the onset of direct martensitic transformation M_s in Ti (Ni, Nb, Mo) alloys, obtained from the temperature curves of martensitic shear stresses. The dashed lines show the region reflecting the growth trend of M_s in relation to the concentration of the alloying element NB.

obtained as a result of this study (Fig. 3.11). To understand the results, we use a phenomenological approach based on the analysis of phase diagrams in the ε–T coordinates (Fig. 3.12). This diagram is schematic and constructed for a single crystal with a flat free planar phase boundary [30]. We assume that at temperatures below T_0 the martensitic phase in an unstressed crystal is equilibrium. Then, near T_0, there are two such temperatures M_s and A_f (where $M_s < T_0 < A_f$); in a real sample M_s corresponds to the temperature of the onset of the forward MT upon cooling, and A_f to the temperature of the reverse MT upon heating. A sample in an unloaded state at a temperature of $T > T_0$ is in an austenitic state (point A in Fig. 3.12).

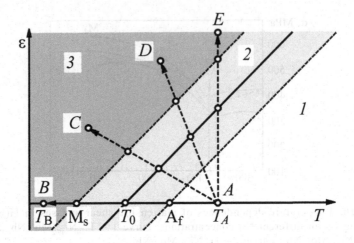

Fig. 3.12. Phase diagram diagram in the σ–*T* coordinates: 1 – region of stable austenite; 2 – pre-transitional region; 3 – region of stable martensite.

As the temperature of the sample decreases to the temperature T_B, martensitic transformation occurs and the alloy completely transforms into the martensitic phase at the temperature M_s, and this martensitic phase represents the cooling-induced martensite. A different MT path can be generated by an external elastic field, which leads to deformation of the alloys. The transition of the alloy from position *A* to position *E*, when an external stress gives rise to an MT, and the martensitic phase in this case is called stress-induced martensite (Fig. 3.12) [6]. With thermomechanical action, i.e., with a simultaneous decrease in temperature and strain, the trajectory of the transition from state *A* (austenitic phase) to state *C* is transformed into the trajectory of the transition from state *A* to *D*. From this diagram it is clear that the thermomechanical effect would lead to an increase in temperature.

It is natural to assume the following [7]: in the region of martensitic transformation, alloys of this class exhibit low-stability pre-transitional states, and the structure and properties of these states depend on the prior history, that is, on previous thermomechanical influences and states of the system. These low-stability states can either alternate with each other or coexist together, forming mixtures of structures. It is logical to expect the following: in a direct martensitic transformation, the set and sequence of the realized low-stability structural states differs from the those of low-stability states in the case of reverse MTs. Consequently, in the direct and reverse transformations, the number of objects scattering the conduction electrons will differ, i.e., as a result of the direct and reverse transformation (one thermal cycling), a hysteresis loop will be observed. With an increase in the number of thermal cycles, the loop width will decrease, which is observed experimentally [8, 9]. Under thermomechanical action, not only does the variety and sequence of low-stability states change, but the set of such states decreases as the transition temperature shifts to the region of higher values.

Conclusion. It has been demonstrated that the temperature of the onset of the direct martensitic transformation M_s, obtained from the temperature dependence of the martensitic shear stresses, has different values and a different functional dependence in comparison with the observed values and dependences observed in unstressed specimens and the temperature dependences of electrical resitivity obtained as a result of this investigation. To understand the results obtained, a phenomenological approach based on the analysis of phase diagrams is used. From the analysis it follows that the thermomechanical effect would lead to an increase in the temperatures M_s and A_f. A logical assumption is made about the relationship between the observed effect and the presence in the alloys of this class of low-stability pre-transitional states in the region of martensitic transformation, and the structure and properties of these states depend on previous thermomechanical influences and system states.

3.3. The effect of phase hardening on pre-martensitic states and martensitic transformations in multicomponent Ti alloys (Ni, Co, Mo) with shape memory effects

The choice of alloys for creating new functional materials with SME is dictated by a number of attractive properties, in particular, the temperature ranges of martensitic transformations, SMEs, martensitic shear stresses, resistance to deformation and temperature cycling, etc. [1–4]. In order to form the optimal physical and mechanical properties of materials, it is necessary to study TiNi-based multicomponent alloys.

Structures made of the TiNi-based alloys are often exposed to external load and temperature during their service. When designing such structures, due attention has to be given to the stability and predictability of the properties of the material. It is known [5] that the critical temperatures and critical stresses for the $B2 \leftrightarrow B19'$

martensitic transformation into TiNi are very sensitive to cycling both in the case of thermal and mechanical cycling. Naturally, during thermal cycling through the region of martensitic transformations in the alloys, i.e. they are subjected to 'phase hardening' [1].

Therefore, to understand the nature of the influence of thermal cycling on the properties of alloys, the information on changes in parameters reflecting the physicomechanical properties of alloys in the MT region is very important.

In the light of the foregoing, the goal is to study the effect of thermal cycling on the physical properties and MTs in multicomponent Ti (Ni, Co, Mo) alloys.

Material and experimental procedure. Six alloys were smelted in an induction furnace. In the preparation of the alloys, the following alloying scheme was used: $Ti_{49.95}Ni_{49.75-x}Mo_{0.31}Co_x$, i.e., Co atoms are replaced by Ni atoms. The cobalt concentration in the alloys was varied from 0 to 3 at.% (Table 3.1). Melting was carried out in an atmosphere of inert argon gas. The resulting ingots were ~1 mm thick, which were then annealed at 850°C for 1 h for homogenization.

The second series of alloys were annealed at a temperature of 450°C for 1 h, followed by quenching in ice water.

Smelting quality was monitored by X-ray diffraction analysis on a Shumadzu XRD-6000 diffractometer. For this experiment, the $Ti_{49.94}Ni_{49.75-x}Co_xMo_{0.31}$ alloys were annealed at 450°C for 1 h, followed by quenching in ice water.

The temperature curves of electrical resistivity were recorded on the laboratory setup using standard four-pin connection of wires to thin specimens.

Martensitic transformations in TiNi-based alloys and temperature dependences of electrical resistivity. A typical method by which the temperature ranges of the direct and reverse MTs and

Table 3.1. The chemical composition of the studied alloys

No.	Concentration of elements in the alloy, at.%			
	Ti	Ni	Mo	Co
1	49.94	49.25	0.31	0.5
2	49.94	48.75	0.31	1
3	49.94	48.25	0.31	1.5
4	49.94	47.75	0.31	2
5	49.94	47.25	0.31	2.5
6	49.94	46.76	0.31	3

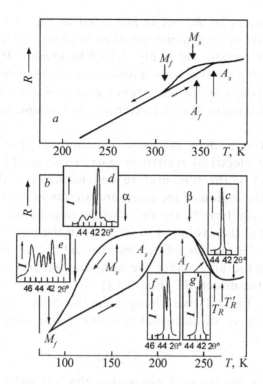

Fig. 3.13. Temperature dependences of electrical resistivity for two MT sequences in TiNi-based alloys: *a*) $B2 \rightarrow B19'$ ($Ti_{50}Ni_{50}$); *b*) $B2 \rightarrow R \rightarrow B19'$ ($Ti_{50}Ni_{47}Fe_3$). The insets (*c–f*) show fragments of X-ray diffraction patterns illustrating the phase compositions at various temperatures.

the sequence of transformations in TiNi-alloys are determined is measuring the electrical resistivity as a function of temperature (Fig. 3.13). Figure 3.13 *a* illustrates the dependence of resistivity on temperature of a Ti–Ni binary alloy hardened by high temperatures. The result of martensitic transformation is the monoclinic phase $B19'$. The $B2 \rightarrow B19'$ transformation is thermoelastic and is accompanied by a gradual growth of martensitic crystals upon cooling during the direct transformation. In the reverse transformation, martensitic crystals decrease in their number and then completely disappear during heating.

The second more complex MT sequence is shown in Fig. 3.13 *b*. The first transformation upon cooling, characterized by a sharp increase in resistivity and a very small temperature hysteresis (1–2 K), reflects the transformation from $B2$ to the rhombohedral R-phase. The temperature corresponding to the beginning of the direct transformation $B2{\to}R$ is indicated in Fig. 3.13 b with the symbol T_R. The second transformation is characterized by a large temperature hysteresis, and in Fig. 3.13 b by symbol M_s corresponds to the beginning of the conversion from R- to $B19'$-phase. The characteristic points on the temperature curves of electrical resistivity allow us to determine the characteristic MT temperatures (T_R, M_s, M_f, A_s, A_f).

In the literature, there are various theoretical approaches to the description of electrical resistivity curves in the MT region [31]. For instance [3], the temperature dependences of the electrical resistivity in the region of the phase transitions in TiNi-based alloys are discussed similarly to the Peierls-type transitions.

We estimated the evolution of changes in the physical properties in the alloy under various influences (thermal cycling and annealing) by measuring the area under the temperature curve of electrical resistivity in the direct MT (Fig. 3.14).

To do so, the relative changes in the peak areas under the temperature curves of electrical resistivity were determined using the expression

$$\delta S = (S_1 - S_i)/S_1,$$

where S_1 and S_i are the peak areas after the 1-st and i-th cycles.

A number of important points should be noted. Firstly, a feature on the temperature curve of electrical resistivity is a consequence of a superposition of a number of phenomena: the appearance of crystals of a new phase, which has different values of resistivity and is characterized by the presence of a high concentration of phase boundaries; the presence of elastic stress fields during the growth of crystals of a new phase and other phenomena. All of them can be represented using a system of phenomenological equations describing the $B2{\to}R{\to}B19'$ sequence [32, 33].

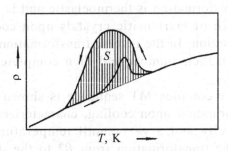

Fig. 3.14. Schematic showing calculation of the area under the resistivity curve.

Secondly, the classical temperature dependence of the electrical resistivity of conductors in the temperature range far from the MT represents a linear temperature dependence:

$$\rho = \rho_0 \, (1 + \alpha_T).$$

The graph is a straight line with a slope coefficient of α. Any deviation of the $\rho = f(T)$ dependence from the straight line in the temperature region preceding the MT indicates a change in the number and type of electron scattering centres commonly representing structural defects, phonons, etc. (for example, large static displacements of atoms from the crystal lattice nodes, i.e., violations of the translational periodicity of the lattice, are some analogues of thermal phonons). It is these significant static displacements that are observed in pre-transitional low-stability states of alloys of this class [34].

Thus, in fact, the area under the curve actually characterizes the 'power' of the formation of defect-scattering conduction electrons. It can be understood that in this low-stability pre-transitional region, the coexistence of a whole set of virtual and real phase and structural states of the system is realized, which are symmetrically different and thermodynamically similar. For example, it was shown in [35] that the interaction of the strain fields developing in the vicinity of defects, depending on their symmetry, contributes to the realization of a martensitic transition from BCC to FCC or ω-like structures. In this approach, a change in the area under the curve characterizes the rate of accumulation of 'defects' scattering the electrons (centres of nucleation of possible phases and structures) on the i-th cycle, i.e. in the thermodynamically low-stability state of the system, numerous possible structures coexists, from which a chain of structures of the actual structural-phase transformation (transition path) is later realized. Naturally, this is an integral characteristic that characterizes the low-stability (in a sense, liquid-like) state of the system and its tendency to the realization of a whole set of different structural states near the stability threshold [35, 36].

In the case under consideration, the thermomechanical treatment of the TiNi alloys leads to the creation of a defective structure, which is well manifested both in a change in the temperature dependences of the electrical resistivity resistivity and in the X-ray diffraction line broadening [35]. Structural defects, due to their elastic stress fields, have a significant effect on pre-martensitic low-stability states due

154

to the short-range order of atomic displacements in phase *B*2 and intermediate shear structures.

Behaviour of electrical resistivity in Ti (Ni, Co, Mo) alloys. As a result of studies of the dependences of the electrical resistivity on temperature, characteristic curves were obtained for alloys with a sequence of $B2{\rightarrow}R{\rightarrow}B19'$ transitions (Fig. 3.15). Their analysis revealed the changes in both the shape of the electrical resistivity curves and the temperature range of features of the concentration dependence of the alloying element Co, where an unusual behaviour is observed. With an increase in the Co concentration, the peaks in the dependences broaden with a decrease in temperature and the characteristic temperatures of the MT shift (dashed lines in Fig. 3.16), while the sequence of the $B2{\rightarrow}R{\rightarrow}B19'$ MTs does not change.

Based on the above temperature behaviour, the concentration dependences of the characteristic temperatures of the martensitic transformation for Ti (Ni, Co, Mo) alloys were constructed (Fig. 3.16).

A comparison of the results obtained with the data of [37, 49] reveals that the temperatures T_R of the onset of $B2{\rightarrow}R$ transformation coincide with the data of other authors for ternary alloys Ti(Ni, Co). In this case, the temperature region of the $R{\rightarrow}B19'$ MT in the

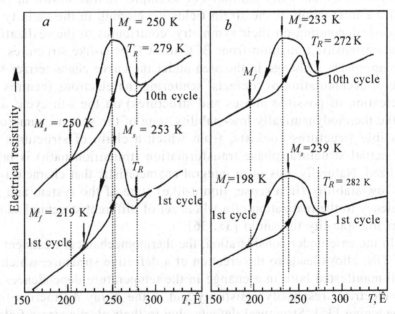

Fig. 3.15. Temperature dependences of the electrical resistivity $\rho(t)$ in the $Ti_{49.94}Ni_{46.76}Co_3Mo_{0.31}$ alloy: *a* – after quenching from 800°C (1 h); *b* – after annealing at 450°C (1 h).

multicomponent alloys under study is noticeably lower than in the ternary alloys Ti(Ni, Co), i.e. doping with Co leads to a considerably lower characteristic MT temperatures than in these TiNi alloys, but at the same time it does not significantly affect the temperature T_R of the $B2 \rightarrow R$ phase transition.

Effect of annealing on martensitic transformations in $Ti_{49.94}Ni_{50-x}Co_xMo_{0.31}$ alloys. A study of the effect of annealing on the temperature dependences of the electrical resistivity revealed the following.

In the specimens annealed at $T = 450°C$ for 1 h and quenched after annealing from $T = 800°C$, the temperatures T_R of the onset of direct $B2 \rightarrow R$ and $R \rightarrow B19'$ MTs remain practically unchanged. At the same time, the shape of the curves of the temperature dependences of the electrical resistivity in the MT region changes differently at different concentrations of the alloying element (Fig. 3.15). The graphs show that in binary and multicomponent alloys with a low content of the alloying element, the peak of the electrical resistivity curve in the MT region becomes sharper after annealing, and in the

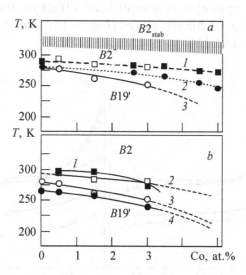

Fig. 3.16. Characteristic temperatures of the martensitic transformation for three-component Ti (Ni, Co) alloys [5] and four-component Ti (Ni, Co, Mo) alloys as a function of the Co concentration: a – after quenching from 800°C according to our data ($\square - T_R$, O – Mn) and the data of [38] (curves 1, 2 – T_R; curve 3 – M_s; O – T_R, ● – M_s); b – after annealing at 450°C (1 h) (O – T_R, ● – M_s) and after quenching from 800°C ($\square - T_R$, O – M_s) (curves 1, 2 – T_R; curves 3, 4 – M_s; O – T_R, ● – M_s).

alloy with 3 at.% Co, there is a noticeable broadening of the peak in the temperature curve of electrical resistivity in the case of a direct MT.

Thus, the annealing was reflected in the temperature dependences of the electrical resistivity as a change only in the shape of the peaks in the region of the MT, but did not lead to a noticeable change in the characteristic MT.

Effect of thermal cycling on the temperature ranges of martensitic transformations in Ti (Ni, Co, Mo) alloys. The effect of thermal cycling on temperature ranges was studied on the temperature dependences of electrical resistivity in multicomponent Ti (Ni, Co, Mo) alloys (Fig. 3.15). These dependences allow us to highlight the following important features. Firstly, thermal cycling led to a slight decrease in the characteristic MT temperatures. Secondly, after thermal cycling, the shape of the peaks of the electrical resistivity curves in the transition region noticeably changed. The dependences of the temperature of the onset of the $R \rightarrow B19'$ MT (M_s) on the cycle number were constructed (Fig. 3.17). These dependences clearly show that M_s reaches saturation after the 10th cycle and thermal cycling slightly lowers the temperature of the onset of MT in the studied alloys. Thermal cycling has a significant effect on the temperature ranges of TiNi-based binary alloys (Fig. 3.18), which is consistent with the known data [5].

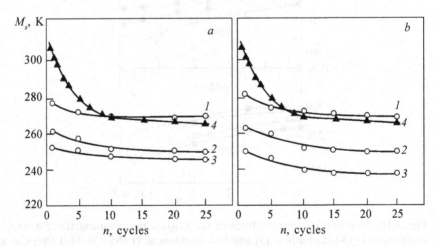

Fig. 3.17. Effect of thermal cycling (heating-cooling-heating) through the temperature range of the MT on the temperature M_s in the Ti (Ni, Co, Mo) alloys quenched from 800°C (*a*) and annealed at 450°C and $t = 1$ h (*b*) : curve 1 – $Ti_{49.94}Ni_{49.25}Co_{0.5}Mo_{0.31}$; curve 2 – $Ti_{48.94}Ni_{48.25}Co_{1.5}Mo_{0.31}$; curve 3 – $Ti_{48.94}Ni_{46.75}Co_3Mo_{0.31}$; curve 4 – $Ti_{49.1}Ni_{50.9}$.

The above dependence of M_s on the cycle number (Fig. 3.17) shows that in the quenched multicomponent Ti(Ni, Co, Mo) alloys, thermal cycling does not have any noticeable effect on the onset temperature of the direct MT (M_s) in contrast with the TiNi-based binary alloys. In this case, the width of the temperature range of the direct MT ($M_s - M_f$) in these alloys monotonously increases with increasing cycle number.

In annealed alloys, the temperatures of the onset of the MT and the nature of the effect of thermal cycling on the MT change (Fig. 3.17 and 3.18). Annealing leads to a noticeable increase in the M_s temperature in the alloys with 0.5 and 1.5 at.% Co, but to a decrease in the alloy with 3 at.% Co (Fig. 3.17).

Thermoelastic MTs clearly exhibit a hysteresis of physico-mechanical properties with a change in temperature in the transition region. The shape and width of the hysteresis loops reflect the features of these transformations [35]. As a result of the analysis of the temperature dependences of the electrical resistivity, the values of the temperature ranges of the direct ($M_s - M_f$) MTs were determined depending on the cycle number. It was found that in quenched alloys thermal cycling leads to a monotonic increase in the MT temperature range with saturation after 10 cycles (Fig. 3.18 *a*). Moreover, the behavior of these curves does not depend on the alloying element concentration, that is, the curves ($M_s - M_f$) = $f(n)$ for alloys with different concentrations of the alloying element

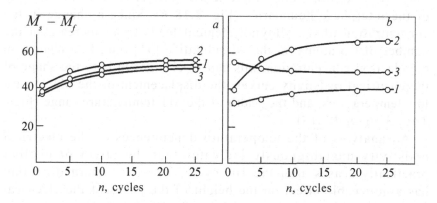

Fig. 3.18. Effect of thermal cycling (heating–cooling–heating) through the MT temperature region on the value of the ($M_s - M_k$) interval in the Ti (Ni, Co, Mo) alloys quenched from 800°C (*a*) and annealed at 450°C for 1 h (*b*) : curve 1 – $Ti_{49.94}Ni_{48.25}Co_{0.5}Mo_{0.31}$; curve 2 – $Ti_{48.94}Ni_{48.25}Co_{1.5}Mo_{0.31}$; curve 3 – $Ti_{48.94}Ni_{46.75}Co_3Mo_{0.31}$.

are similar. A completely different situation is observed for alloys subjected to annealing (Fig. 3.18 *b*). The functional nature of the variation in the $(M_s - M_f)$ as a function of the cycle number (Fig. 3.18 *b*) in alloys with a concentration of 0.5 and 1.5 at.% Co in annealed samples is similar (as in quenched alloys), while in an alloy with 3 at.% Co a decrease $(M_s - M_f)$ is observed as the number of cycles increases.

Such a difference in the $(M_s - M_f) = f(n)$ functional dependences reflects the different effects of thermal cycling on MTs. The following is important. Since the motion of interphase boundaries during the growth of martensitic crystals in the matrix phase leads to the development of elastic stress fields (in this case, the matrix phase is in a low-stability or unstable state), this gives rise to the formation of a wide range of very diverse structural defects (with respect to to the original crystal lattice).

The data obtained allow us to state that in the hardened alloys, thermal cycling monotonically reduces the mobility of martensitic crystal boundaries, regardless of the concentration of the alloying element in the Ti (Ni, Co, Mo) alloys.

Annealing of alloys with different concentrations of the alloying elements leads to different effects of thermal cycling on the mobility of interphase boundaries. In the $Ti_{49.94}Ni_{49.25}Co_{0.5}Mo_{0.31}$ alloy, thermal cycling weakly reduces their mobility (Fig. 3.18, b, curve 1), in the $Ti_{48.94}Ni_{49.25}Co_{1.5}Mo_{0.31}$ alloy it significantly decreases, and in the $Ti_{48.94}Ni_{46.75}Co_3Mo_{0.31}$ alloy there is a slight increase in the mobility of the interphase boundaries (Fig. 3.18 *b*). Since a change in the concentration of the alloying element leads to an increase in the number of Co atoms on the Ni sublattice [31], i.e., to a change in the electron concentration, this causes changes in both the shape of the electrical resistivity curves, the displacement of the MT toward low temperatures, and the width of the MT temperature range (Figs. 3.15, 3.16, and 3.18).

An analysis of the temperature dependences of the electrical resistivity unambiguously indicates that the effect of cycling practically ceases after the 10th cycle. Moreover, thermal cycling has a noticeable effect on the height of the peaks of the electrical resistivity curves in the region of $B2 \rightarrow R \rightarrow B19'$ martensitic transitions.

A further study of the influence of the concentration of the alloying element and heat treatment on the MT was carried out taking into account the changes in the peak areas in the temperature dependences of the electrical resistivity curves in the MT regions. On the dependences of δ_S on the cycle number for annealed alloys, it is easy to see (Fig. 3.19) that, at a low content of the alloying element, there is a significant increase in the peak area on the electrical resistivity curve in the MT region with saturation after the 20th cycle. In alloys with a higher Co content, a slight increase in the area is observed, which is evident in the corresponding curves (Fig. 3.19 a).

Similar dependences were obtained for the annealed alloys (Fig. 3.19 b). Annealing slightly changed the dependence $\delta_S = f(n)$ for the $Ti_{49.94}Ni_{48.25}Co_{0.5}Mo_{0.31}$ alloy (with a low concentration of the alloying element) and practically had no effect in the alloys with a higher content of the alloying element (Fig. 3.19 b, curves 2 and 3). The important role of thermal cycling and heat treatment in the course of MT follows from an analysis of the temperature dependences of the changes in the area under the electrical resistivity curve δ_S in the interval of the direct MT. The dependences shown in Fig. 3.19 allow us to distinguish two groups of alloys. The first group is represented by microalloyed alloys in which the content of the alloying component is ~0.5 at.% Co. Thermal cycling through the MT region in these alloys leads to an insignificant decrease in the MT onset temperature and to a noticeable increase in the area under the curve. Alloys of another group (with a higher Co content from 1.5 to 3 at.%) have a completely different functional dependence $\delta_S = f(M_s)$ (Fig. 3.19). In these alloys, quenched from 800°C, thermal cycling leads to a slight decrease in the temperature M_s and slightly changes δ_S. Thermal cycling of the annealed alloys does not practically

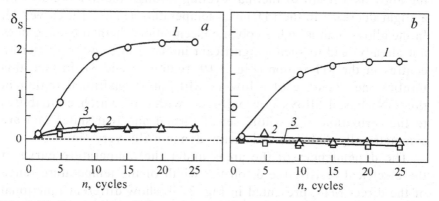

Fig. 3.19. Dependences of δ_S on the temperature of the onset of a direct MT in thermocycled alloys through the MT region: a – quenched from 800°C; b – annealed at 450°C ($t = 1$ h) (curve 1 – $Ti_{49.94}Ni_{49.25}Co_{0.5}Mo_{0.31}$; curve 2 – $Ti_{48.94}Ni_{48.25}Co_{1.5}Mo_{0.31}$; curve 3 – $Ti_{48.94}Ni_{46.75}Co_3Mo_{0.31}$).

change δ_s, but in an alloy with 3 at.% Co, a significant decrease in the temperature of the MT onset is observed. This behaviour reflects structural-phase changes in the alloys after annealing, as a result of which the secondary phases of the type $Ti_3(Ni, Co)_4$, $Ti_2(Ni, Co)_3$, and $Ti(Ni, Co)_3$ have a significant effect on the stability of the austenitic crystal lattice phases with a $B2$ structure. This phenomenon is confirmed in the literature [1–4, 35–46].

Such a different response of the temperature curves of electrical resistivity to thermal cycling is probably caused by the following processes in the alloys. The deviation from the linear temperature dependence of the electrical resistivity in the region preceding the MT is caused by a change in the number and type of electron scattering centres, which can be structural defects or phonons. These defects are accompanied by large static displacements of the atoms from the nodes of the crystal lattice, i.e., the lead to disturbances of the translational periodicity of the lattice. Such significant static displacements are characteristic of alloys in which well-pronounced pre-transitional states are observed before the thermoelastic MTs. Thus, the experimentally discovered dependences of the change in the area under the electrical resistivity curves on temperature in the interval of the direct MT reflect the actually different nature of the change in the 'power' of the formation of defects scattering conduction electrons during thermal cycling as a function of the concentration of the alloying elements.

Microdoping of a TiNi alloy with cobalt was manifested in a well-pronounced increase in the number of electron scattering centres in the alloy as a result of thermal cycling through the MT region with a slight decrease in the MT onset temperature (Fig. 3.19, curves 1). In the alloys with a higher cobalt concentration, thermal cycling does not already lead to such a significant increase in electron scattering centres in the MT region (Fig. 3.19, regions 2 and 3). In fact, this implies the presence of a low-stability pre-transitional region in the TiNi-based alloys microalloyed with Co, which contributes to the formation of a whole set of virtual and real states that are symmetrically different and thermodynamically very similar.

The dependences of the areas under the temperature curve of the electrical resistivitye δ_s on the width of the temperature range of the direct MTs, presented in Fig. 3.29, show different functional dependences $\delta_s = f(M_s - M_f)$ at different concentrations of the alloying element.

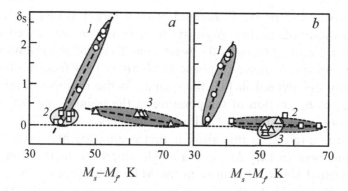

Fig. 3.20. Dependences of δ_S on the width of the temperature range of the direct MT (M_s–M_f) in alloys quenched from 800°C (*a*) and annealed at 450°C (*t* = 1 h) (*b*): cr. 1 – $Ti_{49.94}Ni_{49.25}Co_{0.5}Mo_{0.31}$; curve 2 – $Ti_{48.94}Ni_{48.25}Co_{1.5}Mo_{0.31}$; curve 3 – $Ti_{48.94}Ni_{46.75}Co_3Mo_{0.31}$.

$Ti_{49.94}Ni_{49.25}Co_{0.5}Mo_{0.31}$ **alloy.** In this alloy, microalloying with Co leads to such a state of the crystal lattice of the initial austenitic phase with a *B*2 structure in which thermal cycling causes an intense change in δ_S as a function of (M_s–M_f). This means that the defects arising as a result of phase hardening during thermal cycling hardly affect the mobility of interphase boundaries, but a number of them make a significant contribution to the increase in electron scattering centres. This is also well manifested in the temperature dependences of electrical resistivity. It should be noted that in the TiNi alloys with the same cobalt content, the formation of the *R*-phase during the *B*2→*R*→*B*19′ MT is accompanied by insignificant rhombohedral distortion [4].

$Ti_{48.94}Ni_{48.25}Co_{1.5}Mo_{0.31}$ **alloy.** Thermocycling practically does not change δ_S, however, it has a significant effect on the MT interval: with an increase in the number of cycles, the temperature region of the direct MT increases (Fig. 3.4.8 *a*, region 2).

As a result of annealing, according to [3, 46], secondary phases are formed whose chemical composition differs from the equiatomic one and from the structure of the *B*2-phase. Thus, annealing leads to a new structural phase state that is different from the initial one. Thermal cycling of the alloy in this state has led to a significant broadening of the MT temperature range, but did not change the area under the temperature curve of electrical resistivity δ_S.

$Ti_{48.94}Ni_{46.75}Co_3Mo_{0.31}$ **alloy.** In alloys with a higher content of the alloying element Co (3 at.%), MT *B*2→*R*→*B*19′ occurs with decreasing temperature. A distinctive structural feature is that in this alloy the R-phase is formed with a larger (close to limiting values) rhombohedrality angle α_R than in the microalloyed alloys. The consequence of this is the weak sensitivity of the rhombohedral angle α_R in the R-phase to phase hardening. This feature obeys the regularity discovered in [4]: the change in the rhombohedral lattice angle α_R as a function of temperature in the point T_R decreases with increasing concentration of alloying elements.

What is unusual in this alloy is that during thermal cycling there is an increase in (M_s-M_f) and a slight decrease in the area under the electrical resistivity curve in the MT region (Fig. 3.20 a, region 3). It is known [1–5] that the temperature interval of the MT (M_s-M_f) is associated with the mobility of the interphase boundaries with a change in the phase composition, and the broadening of the temperature region of the MT in this alloy implies that phase hardening during thermal cycling leads to a decrease in the mobility of the interphase boundaries. It can be stated that in the alloys in which the (M_s-M_f) interval has increased significantly, thermal cycling contributed to the formation of defects that had a significant effect on the mobility of interphase boundaries. Annealing at a temperature of 450°C has brought this alloy into such a structural phase state that thermal cycling can no longer changes the width of the MT temperature range (Fig. 3.20 b, region 3).

Conclusion. It has been noted that the nonlinear nature of the change in the temperature dependences of the electrical resistivity under thermal cycling and heat treatment in Ti(Ni, Co, Mo) alloys in the region preceding the martensitic transformation indicates the presence of a region of low-stability pre-transitional states near the stability loss threshold. The unusual effect of thermal cycling through the martensitic transformation region on the temperature dependences of the electrical resistivity curves in the microalloyed multicomponent TiNi-based alloys has been established, namely: thermal cycling leads to a noticeable increase in the area under the temperature curve of electrical resistivity, but no noticeable change in the MT onset temperature has been detected. In Ti (Ni, Co, Mo) alloys with a higher Co content, the character of the change in the curves of electrical resistivity versus temperature differs significantly from similar dependences in the microalloyed alloys. It is important that in the alloys studied in this work thermal cycling ends its significant effect on the MT by the 10th cycle.

The results obtained allow us to argue that thermal cycling through the MT region in alloys with well-defined low-stability pre-martensitic states leads to the development of phase hardening with a wide variety of defects. These defects also affect the mobility of interphase boundaries and can make a significant contribution to the increase in the number of electron scattering centres.

Summary

Using the example of the influence of point defects and their complexes on structural transformations in TiNi-based alloys, it has been shown that in the low-stability state of a condensed system (in TiNi these are the so-called pre-transitional states), the interaction of structural defects can have a significant effect on the features of structural phase transitions, and nanoscale objects play a very important role in the stability of phases in the MT region.

In the pre-transitional low-stability state of a stoichiometric TiNi alloy with the $B2$ superstructure (in the case under consideration, the states with low elastic moduli), significant static displacements of atoms from the sites of the crystal lattice take place before the transformation. As a result, even at a low concentration of defects, they interact with each other. In the final structure, atomic displacements in the vicinity of the defect are localized. Fields of atomic displacements in the vicinity of defects not lying in the planes of the {110} type, prevent shuffling shifts of the {1$\bar{1}$0} planes in the <110> direction and the implementation of the BCC-FCC martensitic transformation by this mechanism. A different variant of the development of martensitic transformation can not be ruled out. Thus, the interaction of defects located in the {111} planes leads to the appearance of a long-period ω-like structure, which is formed by shuffling displacements of the <111> atomic series. Thus, in the pre-martensitic state, when the system is at the threshold of its stability, the interaction of atomic displacement fields developing in the vicinity of defects, can influence the choice of a possible martensitic transition path.

Point defects caused by a deviation from stoichiometry and a disturbance of long-range order in the arrangement of atoms can affect the stability of the $B2$ structure and contribute to the transformation of the martensitic type into a ω-like structure. In this case, high-energy chains of point defects are formed in the <111> direction, which leads to an increase in the number of point defects in the martensitic phase. The latter can be significant in determining the reverse path of the martensitic transition.

Thus, by studying the effect of point defects on structural transformations in TiNi-based alloys, it has been shown that in the low-stability state of a condensed system, the interaction of structural defects can have a significant effect on the structural-phase transformations, and nanoscale objects play a very important role in the stability of the crystal lattice in the region of MT.

When analyzing the effect of deformation on the temperature regions of martensitic transformations in TiNi-based alloys, it was found that the concentration dependences of the M_s temperature of the onset of direct martensitic transformation in stressed and unstressed specimens are different. This made it possible to reasonably assume that the observed effect is associated with the presence of low-stability pre-transitional states in the region of martensitic transformation in the alloys of this class, and the structure and properties of these states depend on previous thermomechanical influences and system states. It was found that the temperature M_s of the onset of the direct martensitic transformation, derived from the temperature dependence of the stresses of the martensitic shift, has a different value and a different functional dependence in comparison with the observed values and dependences in unstressed samples and as a result of studying the temperature dependences of the electrical resistivity. To understand the results obtained, a phenomenological approach based on the analysis of phase diagrams is used. From the analysis it follows that the thermomechanical effect should lead to an increase in the temperatures M_s and A_f. A logical assumption has been made about the relationship between the observed effect and the presence of low-stability pre-transitional states in the region of martensitic transformation in the alloys of this class, and the structure and properties of these states depend on previous thermomechanical influences and system states.

The results of studies of physical properties in multicomponent Ti(Ni, Co, Mo) alloys with the effects of shape memory, the effect of annealing and thermal cycling on the intervals of martensitic transformations and on the pre-martensitic low-stability states are presented. It has been found out that thermal cycling through the MT region in the microalloyed alloys leads to a slight decrease in the MT onset temperature and a noticeable increase in the area under the temperature curve of electrical resistivity with saturation after the 20th cycle. In alloys with a higher Co content, a slight increase in the peak area under the electrical resistivity curve was found. It has been shown that in the $Ti_{48.94}Ni_{48.25}Co_{1.5}Mo_{0.31}$ alloy, the temperature range of the $R \rightarrow B19'$ transformation is independent of thermal cycling. It has been revealed that annealing at a temperature of 450°C leads to a change in the shape of the peaks in the temperature dependences of the electrical resistivity and causes a noticeable change in the characteristic MT temperatures.

It has been found that the concentration dependences of the temperatures of the onset of the direct martensitic transformation M_s in stressed and unstressed specimens differ. It has been presumed that the observed effect is associated with the presence of low-stability pre-transitional states in the region of martensitic transformation in the alloys of this class, and the structure and properties of these states depend on previous thermomechanical influences and system states.

It has been noted that the nonlinear nature of the change in the temperature dependences of the electrical resistivity upon thermal cycling and heat treatment in Ti (Ni, Co, Mo) alloys in the region preceding the martensitic transformation indicates the presence of a region of low-stability pre-transitional states near the stability loss threshold. The unusual effect of thermal cycling through the martensitic transformation region on the temperature dependences of the electrical resistivity in the TiNi-based microalloyed multicomponent alloys has been established, namely: thermal cycling leads to a noticeable increase in the area under the temperature curve of electrical resistivity, but no noticeable change in the MT onset temperature was detected.

In Ti(Ni, Co, Mo) alloys with a higher Co content, the character of the change in the curves of electrical resistivity versus temperature differs significantly from similar dependences in the microalloyed alloys. It is important that in the alloys studied, thermal cycling ends its significant effect on the MT by the 10th cycle.

The results obtained allow us to conclude that thermal cycling through the MT region in alloys with well-defined low-stability pre-martensitic states leads to the development of phase hardening with a wide variety of defects. These defects affect the mobility of interphase boundaries and can also make a significant contribution to the increase in the number of electron scattering centres.

References

1. Gunter V.E., Dombaev G.Ts., Sysolyatin P.G. et al. Medical materials and shape memory implants. - Tomsk: TSU, 1998 .-- 486 p.
2. Klopotov A.A., Gunter V.E., Potekaev A.I. et al. // Izv. Univ. Fizika. - 2009. - V. 52.

166

- No. 9/2. P S. 77–97.

3. Gribov Yu.A., Klopotov A.A., Potekaev A.I. et al. // Izv. Univ. Fizika. - 2010. - V. 53. - No. 1. - P. 65–69.

4. Klopotov A.A., Yasenchuk Yu.F., Potekaev A.I. et al. // Izv. Univ. Fizika - 2008. - V. 51. - No. 3. - P. 7–17.

5. Gunter V.E., Khodorenko V.N., Yasenchuk Yu.F. et al. Nickelide titanium. New generation medical material. - Tomsk: MIC, 2006 .-- 296 p.

6. Gunter V.E., Kotenko V.V., Mirgazizov M.Z. et al.. Alloys with shape memory in medicine. - Tomsk: TSU, 1986.- 206 p.

7. Structure-phase states and properties of metal systems / ed. A.I. Potekaev. - Tomsk: NTL Publishing House, 2004 .-- 356 p.

8. Potekaev A.I., Klopotov A.A., Matyunin A.N. et al. // Material Science. - 2010. - No. 12. - P. 37–44.

9. Klopotov A.A., Gunter V.E., Marchenko E.S. et al. // Izv. Univ. Fizika. - 2009. - V. 52. - No. 11/2. - P. 56–59.

10. Potekaev A.I., Dmitriev S.V., Kulagina V.V. and other weakly stable long-period structures in metallic systems // ed. A.I. Potekaev. - Tomsk: NTL Publishing House, 2010 .-- 308 p.

11. Dmitriev S.V., Medvedev N.N., Potekaev A.I. et al. // Izv. Univ. Fizika. - 2008. - V. 51. - No. 8. - P. 73–79.

12. Potekaev A.I., Kulagina V.V. // Izv. Univ. Fizika. - 2008. - V. 51. - No. 11/3. - P. 148–150.

13. Dmitriev S.V., Potekaev A.I., Nazarov A.A. et al. // Izv. Univ. Fizika. - 2009. - V. 52. - No. 2. - S. 21–26.

14. Potekaev A.I., Dudnik E.A., Starostenkov M.D., Popova L.A. // Izv. Univ. Fizika. - 2008. - V. 51. - No. 10. - P. 53–62.

15. Potekaev A.I., Naumov I.I., Kulagina V.V. et al. Natural long-period nanostructures. - Tomsk: NTL Publishing House, 2002 .- 260 p.

16. Potekaev A.I., Klopotov A.A. Kozlov E.V. et al. Low-stability pre-transitional structures in titanium nickelide. - Tomsk: NTL Publishing House, 2004 .-- 296 p.

17. Potekaev A.I., Kulagina V.V. // Izv. Univ. Fizika. - 2008. - V. 51. - No. 11/3. - P. 148-150.

18. Potekaev A.I., Kulagina V.V. // Izv. Univ. Fizika. - 2009. - V. 52. - No. 8/2. - P. 456–458.

19. Kulagina VV, Dudarev E.F. // Izv. Univ. Fizika. - 2000. - V. 43. - No. 6. - P. 58–63.

20. Kulagina V.V. // Izv. Univ. Fizika. - 2001. - V. 44. - No. 2. - P. 30–39.

21. Mercier O., Melton N., Greemafud G., Hagi J. // J. Appl. Phys. - 1980. - V. 51. - No. 3. - R. 1833–1834.

22. Pushin V.G., Kondratiev V.V., Khachin V.N. Pre-transitional phenomena and martensitic transformations. - Yekaterinburg: AN URO, 1998 .-- 320 p.

23. Krivoglaz M.A., Smirnov A.A. Theory of Ordered Alloys. - M .: MF, 1958.- 388 p.

24. Fan G., Chen W., Yang S., et al. // Acta Met. –2004. - V. 52. - P. 4351–4362.

25. Maeda K. Vitek V., Sutton A.P. // Acta Met. - 1982. - V. 30. - No. 1. - P. 2001–2010.

26. Potekaev A.I., Klopotov A.A., Kulagina V.V., Gunther V.E. // Izv. Univ. Ferr. Metallurgiya. - 2010. - No. 10. - P. 61–67.

27. Kulagina V.V., Eremeev S.V., Potekaev A.I. // Izv. Univ. Fizika. - 2005. - V. 48. - No. 2. - P. 16–23.

28. Yamada Y. // Met. Trans. A. - 1988. - V. 19. - No. 4. - R. 777-783.

29. Abramov V.Ya., Aleksandrova N.M., Borovkov D.V., Khmelevskaya I.Yu. // Materials Science and Engineering. - 2006A. - V. 438-440. - P. 553–557.

30. Boyko V.S., Garber V.S., Kosevich A.M. Reversible plasticity of crystals. - M .: Nauka, 1991 .-- 280 p.

31, Novak V., Sittner P., Dayananda G.N., et al. // Materials Science and Engineer-ing. - 2008. - V. A 481–482. - P. 127–133.

32. Klopotov A.A., Potekaev A.I., Kozlov E.V. // Izv. Univ. Fizika. - 2003. - V. 46. - No. 11. - P. 36–40.

33. Gundyrev V.M., Zel'dovich V.I. // Materials Science and Engineering: A. - 2008. - V. 481–482. - P. 231–234.

34. Otsuka K., Ren X. // Progr. Mat. Sci. - 2005. - V. 50. - P. 511–678.

35. Potekaev A.I., Starenchenko V.A., Kulagina V.V. and other Weak-resistant states of metal systems / ed. A.I. Potekaev. - Tomsk: NTL Publishing House, 2012.– 272 p.

36. Klopotov A.A., Potekaev A.I., Gribov Yu.A., Kozlov E.V. // Izv. Univ. Fizika. - 2003. - V. 46. - No. 7. - P. 54–62.

37. Likhachev V.A., Kuzmin S.L., Kamentsova Z.P. The effect of shape memory. - L .: Leningrad State University, 1987 .-- 216 p.

38. Litvinov V.S., Arkhangelskaya A.A. // Heat treatment and physics of metals. - Sverdlovsk: UC AN SSSR, 1982. - 175 p.

39. State diagrams of double metal systems / ed. N.P. Lyakishev. - M.: Mashinostroenie, 1996–2000. - V. 1–3.

40. Kornilov I.I. Metallides and the interaction between them. - M .: Nauka, 1964 .- 180 s.

41. Hume-Rothery V., Raynor G. The structure of metals and their alloys. - M.: Metal-lurgizdat, 1958.

42. Pearson W. Crystal chemistry and physics of metals and alloys. - M .: Mir, 1977 .-- 420 p.

43. Laves F. // Theory of phases in alloys. - M.: Metallurgiya, 1961. - P. 111–199.

44. Kann R.U., Haazen P. Physical metallurgy. V. 1. - M .: Metallurgiya, 1987. - 640 p.

45. Caciamania G., Ferro R., Ansarab I., Dupin N. // Intermetallics. - 2000. - V. 8. - P. 213–222.

46. Pushin V.G., Kondratiev V.V., Khachin V.N. Pre-transitional phenomena and mar-tensitic transformations. - AN UR.

4

Influence of structural defects, low-stability pre-transitional states and structural-phase transformations on the stability of alloys

The inter-relationship between the point, planar defects and their complexes and low-stability pre-transitional states, phase transitions, structural transformations and stability of alloys with respect to structural-phase transformations in the low-stability state of systems is studied using BCC alloys as examples.

It is shown by the computer simulation methods that in a condensed system in the low-stability state, the cooperative interaction of point and planar defects can lead to their ordered arrangement, and the resulting static displacement fields can both stabilize the $B2$ structure and contribute to its instability and martensitic phase transition. In the presence of a certain type of defects in the $B2$ structure, the latter is unstable to shifts of $\{111\}$ planes along the direction of the $<111>$ type. In the final structure, the displacement fields around the defects are localized, and the defects themselves organically fit into the structure of the formed phase. Structural defects of the parent phase become natural elements of the structure of the final daughter phase. Structural defects in the low-stability state of the parent phase determine, in fact, the structure of the final daughter phase.

It is shown that in the vicinity of structural-phase transformations in CuPd alloys with a content of 40 at.% Pd the low-stability states are realized, which demonstrate a number of anomalous phenomena (e.g., anisotropy of atomic displacements, concentration inhomogeneities, stratification, heterophase fluctuations, nonlinearities in the dependences of the lattice parameters and long-range order parameters, etc.) that prepare the system for transformation.

Using the example of point and planar defects and their complexes, the inheritance of structural defects by the daughter phase during structure-phase transformations in the pre-transitional low-stability state of metal BCC systems is studied.

4.1. The influence of structural defects on the stability of alloys with the $B2$ superstructure

Martensitic transformations are observed in many alloys with a $B2$ structure, among which TiNi-based alloys based on it occupy a special place. With decreasing temperature, they exhibit sequences of $B2 \to B19'$ or $B2 \to R \to B19'$ martensitic transformations depending on the composition and thermomechanical treatment [1–6]. Pre-transitional phenomena and low-stability states in these systems are quite prominent. An anomalous temperature dependence of the elastic constants is observed [7–10], the transverse acoustic phonon branch of TA2 $<\xi\xi0>$ experiences a decrease, and there is a dip at $q \approx 2/3$ [11–14]. It is assumed that the condensation of the mode $q \approx 2/3 <110>$ causes the transformation of the second kind into an incommensurate phase, which is then followed by a first-order transition close to the second order transition, to the rhombohedral R-phase.

Doping of TiNi with iron and cobalt atoms, instead of nickel, or a deviation from the stoichiometric composition towards nickel leads to a change in the martensitic transformation sequence from the $B2 \to B19'$ or $B2 \to R \to B19'$. The authors of [1, 5] believe that this may be due to the formation of long-range order in the arrangement of impurity atoms.

In this work, we study the effect of defects periodically distributed in the $B2$ structure, caused by deviations from stoichiometry and a disturbance of long-range order in the arrangement of atoms, on a possible martensitic transformation. The Ti–Ni system was chosen as a model system.

When studying the effect of structural defects on the stability of alloys in a $B2$ structure, the energy of the system was represented as the sum of the pair and volumetric components

$$U = \frac{1}{2}\sum_{i,j} \varphi(r_{ij}) + U(\Omega),$$

where $U(\Omega)$ is the part of the energy that depends only on the atomic volume Ω; $\varphi(r_{ij})$ is the potential of the pair interatomic interaction. For a two-component alloy

$$\varphi(r_{ij}) = P_i^A P_j^A \varphi_{AA}(r_{ij}) + P_i^B P_j^B \varphi_{BB}(r_{ij}) + (P_i^A P_j^B + P_i^B P_j^A)\varphi_{AB}(r_{ij}).$$

Here P_i^A, P_j^A, P_i^B, P_j^B are the occupation numbers of atoms of type A or B of nodes i or j, determined in the usual way, for example $P_i^A = 1$ if atom A occupies node i, and $P_i^A = 0$ otherwise.

To determine the interaction potentials φ_{AA}, φ_{BB}, φ_{AB}, the empirical potential design scheme for the disordered alloy [15] was used as the basis. The indicated scheme was reformulated for the case of an ordered alloy. Following the scheme, the interaction potentials of the like atoms are determined by the properties of pure elements, such as the C_{11}, C_{12}, C_{44} elastic moduli, binding energy, and equilibrium lattice parameters in the structures inherent in these elements. Upon transition to the alloy, the potentials are renormalized taking into account the charge transfer. The potential of unlike atoms φ_{AB} is determined by the given values of the moduli C' and C_{44} and the binding energy.

When considering the point defects and their complexes, the computational cell having the BCC structure in the initial state was selected in the form of a cube with the edges oriented along the axes of cubic symmetry and periodic boundary conditions were used. The use of the latter for a computational cell containing a defect is equivalent to considering an infinite crystal with a defective superlattice. An assumption is made that the processes of structural changes in all primitive cells of the defect superlattice occur synchronously.

The following types of defects were considered:

1) an anti-structural Ni atom on the Ti sublattice;

2) a defect caused by disordering formed by rearrangement of Ti and Ni atoms on the first coordination sphere (Fig. 4.1 a);

3) an anti-structural Ni atom and an order defect forming a linear chain along the <111> direction (Fig. 4.1 b);

4) an excess Ni atom and an order defect located in the {110} plane (Fig. 4.1 c);

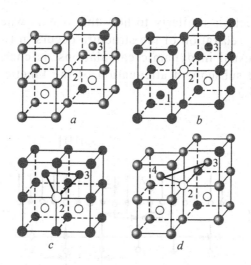

Fig. 4.1. Types of defects considered in $B2$ structure.

5) the same defect, but with the configuration of atoms, shown in Fig. 4.1 d.

When the antistructural nickel atoms are ordered by the sites of the titanium sublattice, the $B2$ structure remains stable. A superlattice of the excess nickel atoms stabilizes the initial structure. The stresses relax only through atomic displacements around the defect. The magnitude of the displacements is significant and, apparently, is associated with the features of the interaction potential obtained using low elastic constants. In cases (2)–(4), the displacement fields near the defects lose cubic symmetry, and contribute to the instability of the $B2$ lattice and the formation of hexagonal martensite through a series of transformations.

Schematically the displacement scheme is shown in Fig. 4.2. The top of the figure shows the packing of the {111} planes in the <111> direction in the $B2$ superstructure. The planes occupied by nickel and tantalum atoms are indicated by black and white circles, respectively. The positions of the planes containing the atoms of the defect are marked with additional circles with numbers. The arrows indicate the direction of the displacements of the {111} planes of the $B2$-lattice. The bottom part of Fig. 4.2 presents the arrangement of the planes of the resulting phase. The lower circles characterize the type of plane, and the upper circles indicate the arrangement of atoms on the <111> line containing defect atoms. It is clear from Fig. 4.2 shows that the {111} planes of the BCC lattice are displaced in the <111> direction similarly to how this occurs when the ω-phase is formed. The relaxation of the structure occurs in two stages, each of which, as usual in dynamic calculations, is characterized by a sharp rise and subsequent attenuation of the kinetic energy of the system (Fig. 4.3).

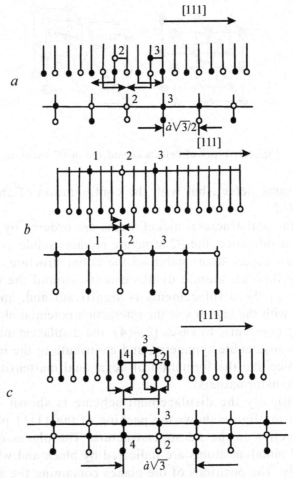

Fig. 4.2. Plan of displacements of planes during transition into the hexagonal phase of the B2 structure containing a defect: a – corresponding to Fig. 4.1 a; b – corresponding to Fig. 4.1, b); c – corresponding to Fig. 4.1 c.

The arrow in Fig. 4.3 indicated the moment of transition from the first stage to the second. This moment corresponds to the formation of an intermediate structure, having the same symmetry as the final one, but atoms of different types are located in the {111} planes.

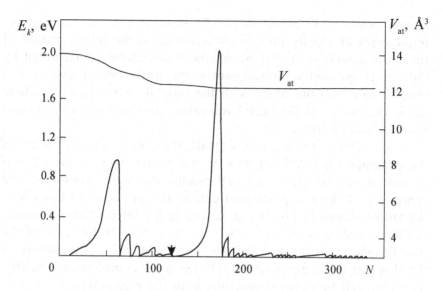

Fig. 4.3. Kinetic energy and atomic volume versus the number of steps N.

The unit cell of the intermediate structure contains 18 atoms. The c axis is directed along <111> and is equal to $a\sqrt{3}$ (a is the lattice constant), the axes a and b equal to $a\sqrt{6}$ are directed along $<\bar{1}\bar{1}2>$ and $<\bar{1}2\bar{1}>$, respectively. With the types of defects, as presented in Fig. 4.2, we get the same final structure, if we do not take into account the fine structure of the defects. The structure is a hexagonal layered ordered phase made of alternating planes containing atoms of the same kind. The layers are perpendicular to the c axis, oriented along the <111> direction of the $B2$-lattice, and the axes a and b are parallel to $<\bar{1}\bar{1}2>$ and $<\bar{1}2\bar{1}>$. The final positions of atoms in the hexagonal structure shown in Fig. 4.2 can be obtained using an equation of the form

$$\boldsymbol{R}_{\mathrm{HP}} = \boldsymbol{R}_{B2} + N[(\boldsymbol{R}_{B2} \cdot \boldsymbol{N}) + A\sin(\boldsymbol{k} \cdot \boldsymbol{R}_{B2} + \varphi)] \quad,$$

where R_{HP} is the vector of hexagonal packing (HP); $\boldsymbol{R}_{B2}$ is the translation vector of the BCC lattice; N is the normalized vector of the normal plane {111}; k is the wave vector; A, φ are the amplitude and phase.

It should be noted that the longitudinal displacement wave does not determine the real trajectory of the atomic motion, but only describes the position of atoms in the martensitic phase using the known positions of atoms in the $B2$ structure. An analysis of the trajectories of atomic motion showed that at the initial instant of time, the direction of displacements is completely determined by the defect symmetry. Subsequently, the trajectory of motion lies approximately in the plane with the normal vector $[\Delta R^x N]$, where ΔR is the vector of the initial relaxation displacement of the atom near the point defect.

If the computational grid is small, then the change in shape of the computational cell depends on the defect type. In the case of an anti-structural atom, the cell retains its cubic symmetry. The symmetry of the computational cell in the presence of defects in the lattice shown in Fig. 4.1 a, b, and in the intermediate structure is rhombohedral, $a = b = c$, $\alpha = \beta = \gamma \cong 93°$. In the case of the transformation with the defect shown in Fig. 4.1 c, the change in the shape of the computational cell was more complicated. Initially, the cubic cell becomes monoclinic with the parameters $a = c \neq b$, $\alpha = \gamma = 90°$, $\beta \approx 99°$, which is consistent with the symmetry of the introduced defect. The formation of such a cell has a corresponding relaxation stage on the curve of the kinetic energy versus the number of time steps.

Therefore, one could have expected the formation of some other metastable structure, but it was not possible to identify it by analyzing atomic displacements. A further relaxation process leads to a rhombohedral structure. Thus, the cubic superlattice of defects after the transition from $B2$ to the hexagonal structure turns out to be rhombohedral. The orientation of the axes of the martensitic phase is related to the symmetry of the defect located in the computational cell. The fields of displacements around the defect in the final structure are localized.

Calculation with the defect shown in Fig. 4.1 d, does not lead to structural transformation. There are significant relaxation-induced distortions near the defect.

The transformation into a hexagonal structure, in addition to the type of defect introduced, is significantly affected by the boundary conditions. A transition to a phase with hexagonal symmetry becomes possible if the computational cell is a multiple of $3N \times 3N \times 3N$, where N is an integer. In this case, an integer number of displacement waves are stacked in the $<111>$ direction, and the boundary conditions do not impede the transfer of the martensitic shift from one cell to another. If these conditions are not met, for example, for the computational cell of $5 \times 5 \times 5$, no structural transition is observed.

The periodic boundary conditions for the simulated defective cell determine the superlattice constructed from these defects. In other words, from the very beginning it is accepted that the ordered arrangement of defects in a crystal is more advantageous. To assess the validity of this assumption, a cell with sizes $3a \times 6a \times 3a$ was considered. If two identical defects are located in such a cell at a distance of $3a$ along the y axis, then it will be similar to the doubled cell $3a \times 3a \times 3a$. Two excess Ni atoms were introduced into the cell $3a \times 6a \times 3a$, which were not bound by the translation vectors. In this case, due to the interaction of the elastic fields of defects, there is a clear tendency to order defects and the formation of a superlattice of defects similar to that for the $3a \times 3a \times 3a$ cells. The transition from the $B2$ structure to the hexagonal one also does not occur either as is the case with the $3a \times 3a \times 3a$ cell for the defect under consideration.

A realistic potential in modelling martensitic transformations would describe a decrease in the moduli with decreasing temperature. There is currently no such potential for TiNi. However, one can trace the dynamics of the pattern of shears caused by the presence of a defect when approaching the phase transformation point using the technique reported in [16–18]. It is possible to construct a series of interaction potentials for various values of elastic constants corresponding to certain temperatures, and already with these potentials to study the lattice relaxation as a function of the magnitude of the elastic moduli. Figure 4.4 shows a diagram of displacements caused by an order defect (Fig. 4.1 a), but for a shear modulus C' four times higher than the previously used value. The nature of the displacements of the planes has not changed – the {111} planes are shifted in the <111> direction. However, the final configurations differ, namely: in this case, the two like planes {111} collapse, one plane remains practically in its place. The atoms along the defect line leave the {111} plane and occupy positions between the planes. The displacements of atoms in the planes themselves, as before, are insignificant.

Despite the fact that the interaction potentials are adapted to the experimental values of the elastic moduli, it was not possible to obtain the {110}<1$\bar{1}$0> shear system characteristic of the martensitic transformation in TiNi alloys.

Thus, in systems with low elastic moduli (this is understood as the condensed system in a low-stability state), long-range displacement fields appear around point defects, therefore, even at a low concentration, the defects turn out to interact with each other. In

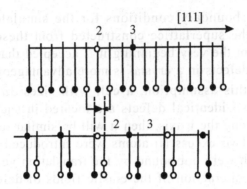

Fig. 4.4. Scheme of displacements of planes at high shear modulus C'

this case, they cannot be treated as isolated defects. The cooperative interaction of point defects can lead to their ordered arrangement, and the resulting static displacement fields can both stabilize the $B2$ structure and contribute to its instability and cause a martensitic phase transition. In the model under consideration, in the presence of a certain type of defects in the $B2$ structure, the latter turned out to be unstable to shifts of the {111} planes along the <111> type direction. In the final structure, the displacement fields around the defects are localized, and the defects themselves organically fit into the structure of the resultant phase. It should be noted that structural defects of the parent phase become natural elements of the structure of the final daughter phase. Structural defects in the low-stability state of the parent phase determine, in fact, the structure of the final daughter phase.

The structure of a point defect inherited by martensite varies depending on the final reaction product. The number of point defects of the order during martensitic transformation, as a rule, increases. High-energy defects are formed from linear chains oriented along the <111>-type direction. These defects favour the reverse martensitic transformation to occur exactly backward.

A change in the number of defects during the martensitic transition leads to the accumulation by the defects of the energy associated with the chemical contribution to the thermodynamic potential, since the number of pairs of unlike atoms in the first coordination spheres changes. The displacements of atoms around defects in martensite are more localized than in the high-temperature phase. Therefore, the elastic energy associated with the non-chemical contribution from the defect decreases during the $B2 \rightarrow \omega$-like martensite transition.

Thus, for the order defects of the anti-structural atoms and their simplest complexes, it is shown that in some cases the number of point defects increases during martensitic transformation. At the same time, the proportion of chemical energy accumulated by defects increases. Some types of defects in the $B2$ structure in martensite have corresponding high-energy linear disordered chains in the <111> type direction. This can play a significant role in determining the path of the reverse martensitic transformation.

Conclusion. It has been shown that in the low-stability state of a condensed system, the cooperative interaction of point defects can lead to their ordered arrangement, and the resulting static displacement fields can both stabilize the $B2$ structure and contribute to its instability and a cause of a martensitic phase transition. In the presence of a certain type of defects in the $B2$ structure, the latter is unstable to shifts of {111} planes along the <111> direction. In the final structure, the displacement fields around the defects are localized, and the defects themselves organically fit into the structure of the resulting phase. It should be noted that structural defects of the parent phase become natural elements of the structure of the final daughter phase. Structural defects in the low-stability state of the phase eventually determine the structure of the final daughter phase.

4.2. The formation of a relaxation-type columnar structure in BCC alloys

Let us try to evaluate the possibility of the formation of ordered low-stability states with a large relaxation-type period in binary alloys with a BCC lattice. By analogy with FCC alloys [19–33], we consider the features of the formation of antiphase long-period relaxation structures based on the $B2$ base superstructure at $T = 0$ K. The relaxation processes stabilizing the long-period state will be simulated by static atomic displacements from ideal positions (lattice modulation).

Since there are no literature data on the experimental observation of such structures, this study will be predictive in nature.

Features of the model of low-stability antiphase columnar structure. Let us consider the possibility of the formation of a low-stability columnar structure as a result of the formation of two systems of periodic intersecting antiphase boundaries (APBs) 1/2 [111] (1$\bar{1}$0) and 1/2 [$\bar{1}\bar{1}$1] (101) in an alloy of equiatomic composition with a basic $B2$ superstructure. The choice of APBs of this orientation is associated with their low energy of formation. In [21], it is indicated that APBs of this type have the least energy among other orientational variants. It can be expected that the increase in the atomic bond energy during the formation of APBs will be small and the relaxation effects in the form of atomic displacements will stabilize the equilibrium APB atoms.

For convenience, we choose a coordinate system whose basis vectors are

$$a = \frac{1}{2}[111]; \quad b = \frac{1}{2}[\bar{1}\bar{1}1]; \quad c = \frac{1}{2}[1\bar{1}\bar{1}]$$

along the x', y', z' directions, respectively.

In the future, we will assume that the alloy consists of identical antiphase domains with dimensions $M_x = n$ along the x' direction and $M_y = m$ along the y' direction. In this case, M_x and M_y are measured in the lattice parameters $a_x = 2a$ and $b_y = 2b$.

In this consideration, we restrict ourselves to the nearest-neighbour interaction, which we will set by the Morse potential function. Since there is no angular dependence in the potential, it is more convenient to convert the oblique coordinate system (x', y', z') into a orthogonal one (x, y, z), while maintaining interatomic distances. If necessary, the results can easily be returned to the system (x', y', z').

It is easy to show that along the z direction we can restrict ourselves to considering two atomic planes, to which we assign the indices $k = 1$ and $k = 2$ (Fig. 4.5). From the symmetry of the domain in the xy plane, it is sufficient to include only half of it along the

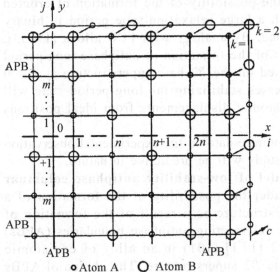

Fig. 4.5. Computational grid scheme.

○ Atom A   ◯ Atom B

x or y axis in the calculated complex. In Fig. 4.5, the calculated complex is limited by APBs, which are marked by the bold type lines, and by half of the domain along the y direction, which is indicated by the dashed line.

Let us assume that in the xy plane, each atom in the node (i, j, k) of the computational grid has individual coordinates $x_{i,j,k}$ and $y_{i,j,k}$, and atomic planes are shifted along the z axis. This assumption is expedient, since it was shown in [21] that only normal atomic displacements are characteristic for a single APB with the (110) plane.

The conditions for the coordinates at the boundaries of the complex under consideration have the form

$$x_{-i,j,k} = -x_{i,j,k}; \; y_{-i,j,k} = -y_{i,j,k}$$

$$x_{2n+l,j,k} = 2x_m - x_{2n+1-l,j,k}; \; y_{2n+l,j,k} = y_{2n+1-l,j,k}$$

$$x_{i,m+h,k} = x_{i,m+1-h,k}; \; y_{i,m+h,k} = 2y_M - y_{i,m+1-h,k}$$

$$x_{i,-j,k} = x_M - x_{2n+1-i,j,k}; \; y_{i,-j,k} = -y_{2n+1-i,j,k},$$

where i, $l = 1, ..., 2n$; j, $h = 1, ..., m$; $k = 1.2$ and x_M, y_M are the coordinates of the antiphase boundaries along the x and y axes, respectively. From the symmetry of the complex under consideration, it is easy to obtain the coordinates of atoms in the $k = 2$ plane from the corresponding values of the $k = 1$ plane:

,

$$x_{i,j,2} = x_M - x_{2n+1-i,j,1}$$

$$y_{i,j,2} = y_{2n+1-i,j,1}.$$

Under the accepted external conditions, the state of the system will be characterized by internal energy, which for the complex under consideration is written in the following form:

$$E = \sum_{i=1}^{2n}\left[\sum_{j=1}^{m-1} W_{i,j,1}^{i,j+1,1} + \frac{(W_{i,1,1}^{i,-1,1} + W_{i,m,1}^{i,m+1,1})}{2}\right] +$$
$$+ \sum_{j=1}^{m}\left[\sum_{i=1}^{2n+1} W_{i,j,1}^{i+1,j,1} + \frac{(W_{1,j,1}^{-1,j,1} + W_{2n,j,1}^{2n+1,j,1})}{2}\right] + \sum_{i=1}^{2n}\sum_{j=1}^{m} W_{i,j,1}^{i,j,2},$$

where $W_{i,j,k}^{l,h,p}$ is the interaction energy of atoms located in the nodes (i, j, k) and (l, h, p).

Then the energy of the defect formed (pairs of intersecting antiphase boundaries) per alloy atom is

$$e' = E/2nm = e_1 - e_0,$$

where e_0 is the internal energy per atom of the alloy with the $B2$ superstructure.

The equilibrium value of the internal energy of the defect per atom of the alloy was found by minimizing e' in $4nm+3$ independent variables: x_M, y_M, c and with respect to $2nm$ variables of types $x_{i,j,k}$ and $y_{i,j,k}$. The variables were changed in such a way that the total energy of the system decreased. Note that in this case, due to the variation in the values of x_M, y_M, and c, a certain change in the volume of the alloy was taken into account upon transition to a long-period state. Pay attention to one more point. Since the atomic compositions of the $k = 1$ and $k = 2$ planes does not differ, the anisotropy of the interatomic interaction will be the leading factor in the formation of a columnar antiphase structure. In fact, this is an extension of the relaxation-type long-period structure (as noted in the model formulation).

Analysis of the possibilities of implementing a low-stability antiphase columnar structure and its structural features. In calculating the internal energy, it was assumed that the interaction of atoms along the z direction differs from that along the x and y axes. This was dictated in the choice of different α and R^0_{AA}. In the calculations we used the potential parameters given in Table 4.1. The values related to the z axis are designated as α (1) and R^0(1). The following values were used in the calculations:

$$D_{AA} = 6.0 \cdot 10^{-19} \text{ J}, \ R^0_{AA} = 2.62 \cdot 10^{-10} \text{ m},$$
$$D_{BB} = 7.5 \cdot 10^{-19} \text{ J}, \ R^0_{BB} = 3.10 \cdot 10^{-10} \text{ m},$$
$$D_{AB} = 7.0 \cdot 10^{-19} \text{ J}, \ R^0_{AB} = 2.75 \cdot 10^{-10} \text{ m},$$
$$\alpha_{AA} = 1.93 \cdot 10^{10} \text{ m}^{-1},$$
$$\alpha_{BB} = 1.35 \cdot 10^{10} \text{ m}^{-1}, \ R^0_{AB} (1) = 2.60 \cdot 10^{-10} \text{ m},$$
$$\alpha_{AB} = 1.25 \cdot 10^{10} \text{ m}^{-1}, \ \alpha(1)_{AB} = 1.60 \cdot 10^{10} \text{ m}^{-1}.$$

The internal energy of an alloy with an initial $B2(e_0)$ superstructure with a given interaction was found by minimizing the alloy energy with respect to the lattice parameter. Thus, the values of e_0 and the corresponding lattice parameter in the selected coordinate system were found.

Table 4.1. Parameters of interaction potentials

Atom pair	$D \cdot 10^{12}$, erg	α, Å^{-1}	α(1), Å^{-1}	R^0, Å	R^0(1), Å
$A-A$	6.0	1.93		2.62	
$B-B$	7.5	1.35		3.10	
$A-B$	7.0	1.25	1.60	2.75	2.60

For an antiphase columnar structure with n, $m = 1, ..., 5$, we calculated the surface $e = e(n, m)$, that is, the internal energy of the defect as a function of the M_x and M_y domain sizes. It turned out that the energy surface of the newly introduced unrelaxed defect is in the range of positive values for any M_x and M_y. Thus, a defect without relaxation effects would not form due to an increase in the binding energy of the system. An analysis of the energy surface of the equilibrium relaxed defect showed a minimum in the values of e for $n = m = 3$ (Fig. 4.6). It is easy to see several antiphase domains with different values of M_x and M_y, which have e values lying in the negative region. This means that a whole set of columnar structures with different M_x and M_y provides the energy advantage of a low-stability long-period state relaxation-type state. In addition, the energy difference between these symmetrically distinct structural long-period states is very small. Therefore, it can be expected that with a small thermal force the system would undergo structural-phase transformations, and at non-zero temperatures, a number of low-stability long-period structures would coexist. This indicates that the long-period columnar structure of the relaxation type will be characterized by low stability.

Thus, the formation of an equilibrium low-stability columnar antiphase structure of a relaxation type is quite possible in terms of energy. A necessary condition for its implementation in alloys with a basic $B2$ superstructure is the interatomic interaction anisotropy.

Since the numerical values of $e = e(n, m)$ in the region of the minimum differ very slightly (on the order of 10^{-14} erg), at finite temperatures we can expect the state of a mixture of domains with some discrete range of sizes M_x and M_y. Moreover, it is easy to see that the variety of long-period states provides their energy advantage in comparison with the initial short-period state, therefore, weak

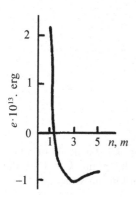

Fig. 4.6. Dependence of the internal energy of APB per atom of an alloy at $n = m$.

external influences can not only transfer the system from one long-period state to another, but such systems would be encountered quite often in systems of a certain class .

The calculation demonstrate that as a result of the introduction of an equilibrium periodic defect, the volume of the alloy changes. In the calculated case, $V/V_0 = 0.9933$, where V is the volume per atom in the final state, and V_0 is its value in the $B2$ structure. In the x and y directions there is an overall expansion of the lattice occurs by 1.2 and 2.0%, respectively, and along the z axis there is a lattice compression by 3%.

As a result of calculations, the individual atomic coordinates were found in the plane $k = 1$. Figure 4.7 shows the image for the cases of $M_x = 3$ and $M_u = 3$. An analysis of the results demonstrates that the nature of the lattice distortions caused by a single APB (for example, periodic APB, normal to the direction x) and the size of the region are in a qualitative agreement with the results of calculating a single APB of this type in the CuZn alloy [25].

Note that along the line of intersection of the antiphase boundaries (Fig. 4.7) there are groups of four atoms A and B. The thin lines in the Fig. 4.2 represent the average positions of the atomic planes, where $\bar{a} = x_M/2n$ and $\bar{b} = y_M/2m$. The largest displacements of atoms (both along the x and y axes) are observed for 'large' B atoms included in the group of four. The 'large' B atoms ($R^0_{BB} > R^0_{AA}$) are displaced from the APBs.

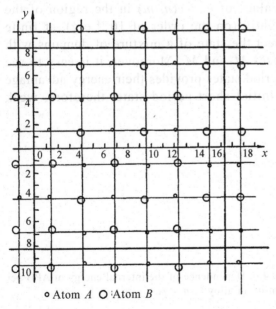

o Atom A O Atom B

Fig. 4.7. Positions of atoms in the plane $k = 1$ of the equilibrium structure.

The 'small' A atoms form groups of four displaced to the APBs. As we move away from the APB, the displacements decrease (Table 4.2). Let us consider an APB normal to the x axis and the displacements of atoms along the x axis. If the 'small' A atom is located at the APB, then this atom and the entire atomic chain along the x direction, are displaced to the APBs. If the 'large' B atom is located near the APB, then the displacements are larger and are directed away from the APB. A similar pattern is observed along the y axis (Table 4.2). In the cubic orientation, the ratios of the lattice parameters of the resulting structure for the calculated case were found to be: $a'/c' = 1.025$ and $b'/c' = 0.996$.

Thermal neutron scattering on a columnar structure. It was determined in the foregoing that a low-stability columnar antiphase structure had been formed by a superposition of APBs intersecting in the columnar structure. For this reason, it makes sense to discuss the features of the diffraction pattern of scattering on a system of parallel periodic equilibrium APBs. On this basis, it is easy to synthesize the resulting scattering pattern on the columnar structure.

In what follows, we will consider the patterns of thermal neutron scattering by an alloy with periodic APBs of the $1/2<111>\{110\}$ type. Following [19–21, 35–39], in this case it is reasonable to study the features in the reciprocal space along the $<111>$ directions near the 110 superstructure reflection of the initial superstructure.

Alloys with a BCC structure often consist of atoms neighbouring in the periodic system of elements. In these cases, it is difficult or impossible to examine the structure by the methods of electron microscopy or X-ray diffraction. The neutron diffraction method is most acceptable in this situation.

Table 4.2. Displacements of atoms from the middle positions ($\Delta x_{ij} = \bar{x}_i - x_{ij}$ and $\Delta y_{ij} = \bar{y}_i - x_{ij}$) of the equilibrium domain $n = m = 3$ in the plane $k = 1$

i	$j=1$		$j=2$		$j=3$	
	Δx, Å	Δy, Å	Δx, Å	Δy, Å	Δx, Å	Δy, Å
1	−0.065	0.070	0.158	0.059	0.062	0.060
2	0.093	−0.109	0.105	−0.145	−0.084	−0.171
3	−0.120	0.078	0.060	0.066	0.108	0, 063
4	−0.146	−0.110	0.034	−0.140	−0.135	−0.170
5	−0.168	0.075	0.028	0.060	0.160	0.061
6	−0.181	−0.098	0.049	−0.132	−0.179	−0.168

184

Thus, we are going to discuss the features of the thermal neutron scattering pattern along <111> near the 110 superstructural reflection of the initial superstructure.

The structural factor of the basis in this case can be written as

$$f(k) = \sum_{k=1,2} \sum_{j=1}^{4m} \sum_{i=1}^{4n} F_{ijk} \exp(-ir'_{ijk}k),$$

where F_{ijk} is the form factor of the atom located in the (ijk) node; $r'_{ijk} = \hat{A} \cdot r_{ijk}$ is the radius vector of this atom in the (x', y', z') coordinate system; $\hat{A}$ is the conversion operator.

The scattering intensity has the form

$$J(k) = f(k)f^*(k).$$

In the calculations, we take the following values of form factors:

$$F_A = 1.03 \cdot 10^{-14} \text{ m and } F_B = -0.34 \cdot 10^{-14} \text{ m},$$

taking the TiNi alloy as an analogue.

Figure 4.8 shows the calculation results for the [111] direction in the case of a columnar structure with dimensions $M_x = M_y = 3$. Here k_0 implies the position of the superstructure maximum of the initial $B2$ superstructure. It is easy to see that with the introduction of a defect, the shape of the curve transforms with the complete disappearance of the superstructural reflection.

In the vicinity of k_0, a set of maxima is formed, and the value of k_0 itself corresponds to a minimum of intensity. The maxima are symmetric with respect to k_0 in position, but asymmetric in intensity.

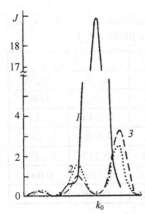

Fig. 4.8. Intensity distribution in the [111] direction in the vicinity of the position of the superstructure reflection from the $B2$ (k_0) superstructure: curve 1 – superstructure peak; curve 2 – with a newly introduced defect $M_x = M_y = 3$; curve 3 – with an equilibrium defect $M_x = M_y = 3$.

In the presence of static displacements in the alloy, which ensure the energy advantage of the columnar structure over the initial $B2$ superstructure, there is an additional redistribution of intensity. The position of the maxima does not change, but the asymmetry of the intensity of reflections increases.

Let us extrapolate to the case of a columnar antiphase structure. In the scattering pattern near the superstructure reflection of the initial $B2$ superstructure, one can expect a certain superposition of the intensity distribution for the two <111> directions, which involving the situation described above. The final form will acquire a cruciform shape with the presence of asymmetry in the peak intensity.

Conclusion. It has been shown that the formation of a low-stability equilibrium columnar antiphase structure of a relaxation type in BCC systems is quite possible as far as the system's energy is concerned. A necessary condition for its implementation in alloys with a basic $B2$ superstructure is the presence of anisotropy (for example, interatomic interaction). Moreover, the columnar structures of various sizes provides the energy advantage of a low-stability long-period state of relaxation type. It can be expected that under a small thermal impact the system will undergo structural-phase transformations, and at non-zero temperatures, a certain number of low-stability long-period states will coexist.

4.3. Low-stability pre-transitional states, order–disorder phase transitions, and $B2$–$A1$ structural transformations in Cu–40 at.% Pd alloys

Recently, a good deal of attention has been given to the low-stability states in condensed systems, which is determined by their structural features and the potential for use as functional materials [21–23, 29, 30, 40]. However, the understanding of the physics of the formation of such states and the features of their structure and properties is still lacking. From this perspective, we believe that the Cu–Pd system, which has been studied for a long time and provided a wealth of the experimental material (for example, [41–45]) is a convenient object of research.

Significant progress has been made at present in understanding the development of phase transitions in Cu–Pd-based alloys (which is especially interesting in the region of 40 at.% Pd) when studied by electron microscopy [46–49] and X-ray diffraction *in situ* [50–52] as well as investigating its physical properties [53, 54]. Despite this, the features of structural changes in the region near phase transitions remain unclear.

Note the experimental facts that have been reliably established in these alloys. First, the order–disorder phase transition (PT) in these alloys is accompanied by the $B2$–$A1$ structural transformation (ordered BCC–disordered FCC). Second, the maximum of the dome, which is characteristic of the order–disorder PT, on the phase diagram is characteristic not for the equiatomic composition, but for the 40 at.% Pd region [55]. This immediately makes the question of the influence of non-stoichiometric composition on the ordering processes in phase order-disorder phase transition important. Third, the state diagram [55] is a conditionally equilibrium diagram.

In CuPd alloys, phase $A1$ can be formed in the low-temperature region by quenching CuPd specimens from temperatures above T_f. This indicates that the $A1$ phase is in a metastable state. On the other hand, deformation at room temperature initiates the $B2{\rightarrow}A1$ phase transition, which implies the stability of the defective $A1$ phase and the low stability or instability of the defective $B2$ phase at low temperatures [52]. This is due to the nature of the existence of stable and metastable phases in the low-temperature region where thermally activated processes have practically no effect on the change in the structural phase state. Naturally, under these conditions it is impossible for the alloy to achieve equilibrium states of the alloy. Fourthly, in the Cu–Pd alloys, in addition to the $B2$–$A1$ phase transition, there are some features related to the unusual behaviour of the crystal lattice both in the transition region [51] and in the pre-transitional region of low stability. The order–disorder PT is controlled by diffusion processes. In this case, the role of low stability or lattice instability is completely unclear. All this indicates that the $B2$–$A1$ phase transition in CuPd alloys is inherently a complex phenomenon.

The task is to reveal the features of structural changes in the region of phase transitions in alloys of the Cu–Pd system in the range of the concentration of ~40 at.% Pd.

Materials and experimental technique. The Cu–39 at.% Pd and Cu–36 at.% Pd alloys were smelted from electrolytic copper and palladium with a purity of 99.99% in an argon atmosphere. Melting was carried out in an inert atmosphere in a resistance furnace. For X-ray studies, both powdered and bulk specimens were used. The disordered state was achieved by quenching from 800°C, and the ordered state was achieved by long-term annealing according to the stepwise cooling mode from 600 to 300°C in increments of 10°C per day. The temperature was maintained with an accuracy of ±2.5°C. The measurements were accompanied by holding the specimens at isothermal temperatures. The equilibrium state was determined out by measuring various parameters (line the intensity ratios of two phases, the ratio of the superstructure lines intensities to the main ones, the angular position of the reflections and their half-widths) as a function of the holding time at different temperatures.

B2 phase transitions. Change in the structure-phase state during transitions. In the course of stepwise heating at a temperature of 570°C in the ordered Cu–39.5 at.% Pd alloy, the $B2 \rightarrow A1$ phase transition begins in the alloy (Fig. 4.9 a, curve 1). With a further increase in temperature, the disordering of the $B2$ phase occurs only at 600°C (Fig. 4.9 a, curve 2).

Figure 4.9 b shows the temperature dependences of the atomic long-range order parameter and the volume fraction of the $B2$ phase in the $A1 \rightarrow B2$ reverse transition. It was found that in this case an ordered phase with a $B2$ structure is formed at 580°C. The degree of atomic long-range order in the $B2$ phase increases in a stepwise fashion in a narrow temperature range ($\sim$10°C) to high values (0.8) of the long-range order parameter. A further decrease in the annealing temperature from 580°C does not give a noticeable increase in the order parameter, while the volume fraction of the ordered phase monotonically increases with decreasing temperature.

The $B2$ and $A1$ two-phase regions do not coincide, which indicates the presence of a hysteresis during the $B2 \rightarrow A1$ transition and corresponds to the real situation in which the transformation occurs with some supercooling below T_0 (or overheating above T_0), necessary for the nucleation of a new phase. From the analysis of the dependences of the change in the volume fraction of the $B2$ phase, it follows that the beginning of the $A1 \rightarrow B2$ transition coincides that of the $B2 \rightarrow A1$ transition (Fig. 4.9 b).

In the CuPd alloy under study, the order–disorder phase transition is accompanied by a $B2 \rightarrow A1$ structural phase transition. Comparing the $\eta = \eta(T)$ dependence in the Cu–39.5 at.% Pd alloy (Fig. 4.10) with the known dependences of the CuZn, FeCo, and AgZn alloys [56], which experience an order–disorder phase transition (second-order phase transition ($B2$–$A2$ PT), it is easy to come to the following conclusion. In the Cu–39.5 at.% Pd alloy, the temperature dependence of the long-range order parameter during heating does not coincide

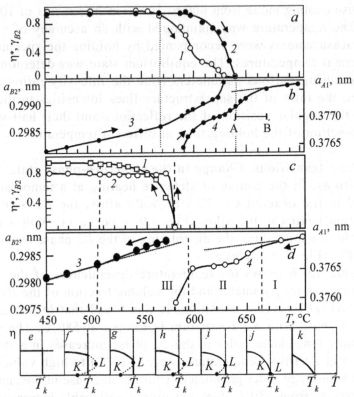

Fig. 4.9. Temperature dependences of the volume fraction of phase $B2$ (curve 1), degree of order in phase $B2$ (curve 2), unit cell parameters in phases $B2$ (curve 3) and $A1$ (curve 4) with phase transition $B2{\to}A1$ (a, b) and $A1{\to}B2$ (c, d) in the Cu–39.5 at.% Pd. PT Classification taking into account the behaviour of long-range order parameters: e – phase I of the first kind; f – order–disorder PT of the first kind; g — critical disordering of the first kind; h – critical ordering of the first kind; i – critical PT of the first kind; j – anomalous PT of the first kind; k – PT of the second kind [56].

with the dependence $\eta = \eta(T)$ for the alloys that undergo only order–disorder PT and whose AB composition is close to the stoichiometry. The curve of the temperature dependence of the atomic long-range order parameter in a phase with the $B2$ structure of the studied alloy is much steeper than similar dependences in CuZn, FeCo, and AgZn alloys (Fig. 4.10). The data presented indicate, on the one hand, a significant difference in the disordering processes in the $B2$ phase during the $B2{\to}A2$ order–disorder phase transition and the structural $B2{\to}A1$ transition. This is due to the presence of two phases in the phase transition region, which adds features of a first order phase transition to this one.

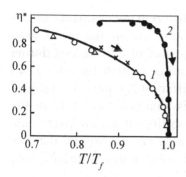

Fig. 4.10. Dependence of the behaviour of the long-range order parameter in the $B2$ superstructure (edge 1) during the order–disorder ($B2$–$A2$) phase transition in AgZn, FeCo, CuZn alloys [28] and long-range order parameter in the $B2$ superstructure at the order–disorder ($B2$–$A1$) transition in the Cu–39.5 at% Pd alloy (curve 2) on the reduced temperature (o – AgZn, × – FeCo, Δ – CuZn).

On the other hand, a change in the order parameter in the $B2$ superstructure in the temperature region close to T_f has certain features that characterize the order–disorder PT as a transition close to type II.

It was shown in [56] that, for the analysis of order–disorder phase transitions, the generally accepted Ehrenfest classification of phase transitions [57], where there only type I and type II transitions must be supplemented with more detailed information (Fig. 4.9, e–k). From the above data, it follows that for the analysis of an order–disorder PT complicated by a structural transition; the behaviour of the $\eta = \eta(T)$ curve in the vicinity of the phase transition should be based on an extended classification [56].

For direct and reverse order–disorder phase transitions, the $\eta = \eta(T)$ dependences in the $B2$ phase are very different (Fig. 4.9 a, b). Upon heating during the $B2 \rightarrow A1$ phase transition, the disordering process in the $B2$ phase is realized by a phase transition close to type II. According to the classification in [99], $\eta = \eta(T)$ belongs to the sixth type, and the phase transition is an anomalous first-order phase transition (Fig. 4.9 a and h). However, upon cooling during the $A1 \rightarrow B2$ phase transition, there is an abrupt order in the order parameter η in the $B2$ phase in a narrow temperature range. This indicates that this transformation is a critical ordering of the first kind and, according to [56], belongs to the fourth type (Fig. 4.9 f). The revealed difference in the types of transitions in the course of ordering and disordering indicates that the reciprocal arrangement of the stability limits of the ordered and disordered phases can lead to a different type of phase transition. In an order–disorder phase transition, not complicated by a structural transition, no such difference was found [56]. This is possible in the case of an order–disorder phase transition occurring simultaneously with the BCC–FCC structural phase transition. This allows us to make an assumption about the existence of different mechanisms of the direct and reverse $B2–A1$ phase transitions in this region of low-stability structural states. Indeed, the existence of a hysteresis during the $B2–A1$ structural phase transition suggests that the processes of ordering and disordering occur in different temperature ranges. In this case, the presence of a hysteresis changes the nature of the phase transition in this region of low-stability structural states, and based on the obtained experimental data, it can be argued that the disordering process in the $B2$ phase is displaced due to the $B2{\to}A1$ phase transition.

Behaviour of the crystal lattice during the $B2{\to}A1$ phase transformation. Figure 4.11 shows the temperature dependence of the lattice parameter of an ordered phase with the $B2$ structure formed by heating the Cu–39.5 at.% Pd alloy from a fully ordered state. The dependence has a linear form up to the PT onset temperature at which an inflection point is observed on the curve. It is easy to see that there are no peculiarities in the behaviour of the lattice parameter before the start of the transformation nor in the temperature dependences of the intensities of the Bragg reflections of the $B2$ structure. Naturally, the temperature dependences of the coefficient of thermal expansion in phases with the $B2$ and $A1$ structures make it possible to reveal features in the temperature dependences of the corresponding lattice parameters. Where there is a kink on the temperature dependence of the lattice parameter, a finite abrupt increase in the coefficient α takes place on the curve of the coefficient of thermal expansion versus temperature. In turn, The finite abrupt increase in the coefficient α shows that in its turn the crystal lattice of a phase with a $B2$ structure in this temperature range has transferred to a new low-stability (less stable) state with respect to the low temperature state.

This indicates the preparation of the crystal lattice for phase transition. In the temperature region, after the $B2{\to}A1$ transition is completed, a feature on the curve is observed, which consists in the deviation of the curve on the linear one.

Behaviour of the crystal lattice during the $A1{\to}B2$ structural phase transformation. We studied the change in the lattice parameter of a phase with structure $A1$ during slow cooling from a single-phase state (Fig. 4.11). In this case, the $A1{\to}B2$ phase transition begins at a temperature of ~590°C. Three sections can be distinguished in the temperature dependence of the lattice parameter. The first is observed far from the transition, on which the dependence is linear. The second is the temperature region immediately before the $A1{\rightarrow}B2$ transition, in which an anomalous temperature dependence is already observed. The third section is the temperature region inside the $A1{\rightarrow}B2$ phase transformation, i.e. the area in which the reflections of a new phase are detected on X-ray diffraction patterns.

Let us address the dependence of the variation in the lattice parameter of the $B2$ ordered phase upon cooling (Fig. 4.12 a). The $A1{\rightarrow}B2$ structural phase transition in this case is completed at a temperature of 510°C. After its completion, the lattice parameter of the $B2$ phase decreases linearly with decreasing temperature; note that is behaviour was similar before the $B2{\rightarrow}A1$ transition (Fig. 4.11 b).

A study of the temperature dependences of the logarithm of the intensity of the structural lines showed that with increasing temperature in the region from 25 to 570°C preceding the $B2{\rightarrow}A1$ phase transition, a linear decrease in the intensity of Bragg reflections of the ordered $B2$ phase is observed. The Debye–Waller factor was determined in the framework of the Debye model which linearly depends on temperature both far from the phase transition and in its vicinity. According to the temperature dependences of the integrated intensity of reflections and the Debye–Waller factor in the ordered phase, there are no signs of instability before the $B2{\rightarrow}A1$ transformation. This indicates a low phase stability. With a further

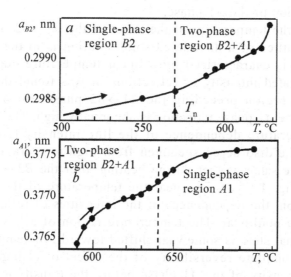

Fig. 4.11. Dependences of the lattice parameters in $B2$ (a) and $A1$ (b) structures during heating from an ordered state during the $B2{\rightarrow}A1$ phase transition.

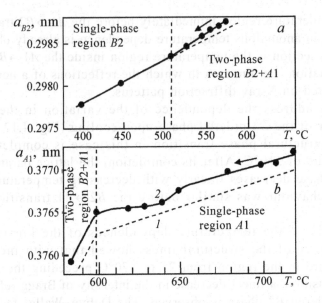

Fig. 4.12. Dependences of the lattice parameters in structures the *B2* (*a*) and *A*1 (*b*): curve 1 – theoretical; curve 2 – experimental during cooling in the course of the *A*1→*B2* phase transition.

increase in temperature, the *B2*→*A*1 phase transition occurs through the two-phase region. An increase in the intensity of reflections of the *A*1 structure in the two-phase region is due to an increase in the volume fraction of the *A*1 phase.

Far from the temperature of the onset of the *A*1→*B2* transition, the crystal lattice of a disordered solid solution is in the ordinary state, which is characterized by a linear temperature dependence of the integrated intensity of reflections. It was found that in the temperature region preceding the *A*1→*B2* phase transition, the intensity of the *A*1 phase reflection is abnormally high. The character of the temperature dependence of the line intensity (331) upon cooling of the disordered specimen from high temperatures (Fig. 4.12 *b*) is the same as that during heating after the *B2*→*A*1 phase transition (Fig. 4.12 *a*), i.e., near the temperature of the onset of transformation, the dependences of the intensities of reflections of phase *A*1 are nonlinear. The temperature region of a low-stability pre-transitional state is somewhat shifted toward low temperatures. There is a complete reversibility of the effect of changes in the reflection intensity of the *A*1 phase before the transition. A slight hysteresis of this phenomenon is probably due to the hysteresis of the *B2*→*A*1 transition.

Thus, in the temperature region of low-stability states preceding the $A1$–$B2$ phase transition, the crystal lattice of $A1$ phase exhibits features that are corresonding to the behaviour of parameters describing the state of the crystal lattice.

Structure-phase transitions $A1_{quench} \to B2 + A1 \to A1$. Variation in the structure-phase state during the PTs. It was established that upon heating of the Cu–36 at.% Pd alloy manufactured by quenching from temperatures above T_f, a change in the structural phase composition is observed: $A1_{quench} \to A1 + B2 \to A1$ (Fig. 4.13 a). In the temperature range 300–600°C $A1$ phase shows a significant concentration heterogeneity, which manifests itself in a noticeable broadening of the reflections of $A1$ phase.

The decrease in the effective long-range atomic order parameter η^* over time corresponds the process of atomic disordering (Fig. 4.13 b). The largest decrease in η^* occurs at the initial instant of time. Further annealing at this temperature slightly changes the order parameter. This is the limiting temperature. With an increase in temperature by 5°C, no phase with the $B2$ structure was detected. The ordered phase $B2$ disappears with a jump in the order parameter ($\eta^* \sim 0.3$).

Peculiarities of the behaviour of the crystal lattice in phases with $A1_{quench} \to B2 + A1 \to A1$ phase transition. Using the X-ray methods, we studied the temperature dependences of the Bragg reflection intensities of the $A1$ structure during heating from low temperatures, formed by quenching from the temperature region above T_f. This condition is metastable. Noteworthy is the different behaviour of the line intensities of reflections obtained from the $A1$ phase as a function of temperature. The intensity of the (111)

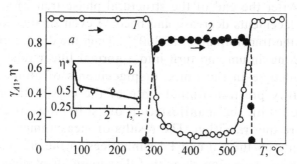

Fig. 4.13. Temperature dependences of the atomic fraction of the $A1$ phase (a, curve 1) and the long-range order parameter in the $B2$ phase (a, curve 2). Dependence of η^* in $B2$ on the time of isothermal annealing at 580°C (b).

reflection decreases before the onset of the phase transition in the temperature range 200–350°C. For the (200), (220), and (331) reflections, a linear dependence is preserved.

Using the X-ray diffraction method, it was possible to detect features on the temperature dependences of the intensities of the diffraction maxima of the disordered phase with the $A1$ structure in both the high-temperature and low-temperature regions before the phase transition. This indicates the existence of a pronounced low-stability pre-transitional state in a disordered solid solution based on the FCC lattice.

Pre-transitional phenomena and low-stability states were commonly observed and investigated in detail in alloys undergoing MTs [21, 22, 32, 36, 37, 40]. In this work, similar phenomena and states were formed in Cu–Pd based alloys with the $B2$–$A1$ structural transition, which occurs simultaneously with the order–disorder phase transition. On the whole, this indicates the diversity of pre-transitional phenomena and the universality of low-stability states, as well as the existence in each particular alloy of the change in structural-phase states before and during phase transitions.

Based on the data obtained, the total r.m.s (root-mean-square) displacements of the atoms from the equilibrium positions were calculated. In the disordered $A1$ phase, fixed by quenching, An abnormal increase in the total number of r.m.s. atomic displacements from the sites of the middle lattice of the disordered phase A1 occurs long before the onset of monotropic (irreversible) phase transition (~150°C). An abnormal change in the total r.m.s. displacements of atoms is also observed in the high-temperature region of the existence of the phase with the $A1$ structure after the completion of the $B2{\rightarrow}A1$ transition. After the end of the structural phase transition, the total atomic displacements decrease and become even smaller than before the start of the transformation at 300°C. Then the displacements pass through the maximum and turn into a normal linear relationship.

Such a change in the r.m.s. displacements of atoms indicates a low-stability pre-transitional state of the crystal lattice, which is associated with the manifestation of 'softening' of the crystal lattice before the transition. The results of measurements of atomic displacements on the Cu and Pd sublattices in the $B2$ phase in the two-phase $B2 + A1$ region show that the atomic displacements in the ordered $B2$ phase have higher displacements on the copper sublattice are larger than the average atomic displacements in the $A1$ phase, but smaller than those on the Pd sublattices .

Thus, the values of the r.m.s. displacements of atoms in the disordered phase with the $A1$ structure, obtained on the basis of experimental X-ray diffraction data on the intensities of Bragg reflections, also make it possible to reveal low-stability pre-transitional states of the crystal lattice in the temperature regions before the start of the monotropic phase transition $A1_{quench} \rightarrow B2$ and after the $B2 \rightarrow A1$ transition. Low-stability pre-transitional states in the low-temperature region significantly differ from those in the high-temperature region after the end of the $B2 \rightarrow A1$ phase transition.

The contribution from the static displacements was isolated from the temperature dependence of the total displacements in a disordered phase with the $A1$ structure. The value of the root-mean-square static displacements of atoms is (0.005 ± 0.003) nm.

The root-mean-square static displacements of atoms in the solid solution were estimated from the FCC lattice of the Cu–36 at.% Pd alloy, the obtained being 0.004 nm. These displacements in a disordered phase with the $A1$ structure make a small contribution to the total r.m.s. displacements in the high-temperature region, and the dynamic r.m.s. atomic displacements make the main contribution. It is worth noting a very important detail: in the temperature regions before the transition in a low-stability state and in the phase transition region, it is practically impossible to differentiate between the static and dynamic contributions by the X-ray diffraction method.

For Cu–40 at.% Pd alloys, the lattice deformation was calculated for the BCC–FCC transition according to the Bain scheme and it was found that its minimum value is ~11%, and the maximum value is 26%. The minimum strain during the rearrangement of the BCC–FCC lattice according to this scheme is comparable with the value of the displacements of the atoms. This suggests that homogeneous Bain deformation along the <110> FCC directions is favourable for the transition to the BCC lattice. The data obtained indicate that the FCC–BCC transition according to the Bain scheme via the homogeneous deformation is difficult. Other mechanisms may more preferable. For example, the FCC–BCC lattice rearrangement can take place according to the mechanism of the short-range displacement-controlled ordering in the FCC structure in its low-stability state regions, using a combination of shear waves with the $k \| <110>$ vector and the longitudinal coordinate along the <110> direction.

Thus, an extraordinary low-stability state of the crystal lattice of a disordered phase with the $A1$ structure before the phase transition is revealed both in the high-temperature equilibrium region of its existence and in the low-temperature metastable region. Such a behaviour of atomic displacements in the $A1$ phase may indicate the development of heterophase fluctuations in a low-stability pre-transitional region.

Concentration inhomogeneities and delamination in CuPd alloys in the region of 40 at.% Pd. Based on the XRD data on the phase composition and concentration variations in the coexisting phases in the two-phase region, a section of the phase diagram of the Cu–Pd system was constructed in the region of 40 at.% Pd. The data on the concentration variations in the phases were obtained from the experimentally determined temperature dependences of the parameters of the unit cells of structures $A1$ and $B2$ in the temperature range 20–700°C and the calculation of the dependences of these parameters on the concentration in alloys of the Cu–Pd system [50].

Figure 4.14 shows the literature data illustrating a part of the equilibrium phase diagram of the Cu–Pd system [55] (Fig. 4.14 a) and the data obtained in the study of structural-phase states in Cu–39.5 at.% Pd and Cu–36 at.% alloys [48].

Let us single out a number of features in the structural phase composition that were revealed during heating of the disordered Cu–36 at.% Pd alloy as a result of the $A1_{quench} \rightarrow B2 \rightarrow A1$ phase transition (Fig. 4.14 b, curve 1). It was found that the half-width of the $B2$ phase reflections practically does not change. It follows that the ordered phase in this transition, in the case of a concentration inhomogeneity, is insignificant. In a disordered solid solution, the equilibrium concentration is reached only at a temperature of 500°C, i.e. in the temperature range from 300 to 500°C, the $A1$ structure has a nonequilibrium concentration. In the same temperature range, the $A1$ structure has a significant concentration inhomogeneity, which manifests itself in the broadening of Bragg reflections with increasing temperature (Fig. 4.14 c). The concentration inhomogeneity sharply decreases only at the moment of reaching the equilibrium state, which coincides with the phase diagram (Fig. 4.14 b), .

The results of studying the concentration change in the disordered and ordered phases, obtained by heating the ordered Cu–39.5 at.% Pd alloy, are presented by the curves in Figs. 4.12 and 4.13, as well as in Fig. 4.14 b, respectively. It was found that the two-phase region ($A1 + B2$) is narrow and there is no broadening of the Bragg reflections of the $A1$ and $B2$ phases in it, which indicates the absence of significant concentration inhomogeneity.

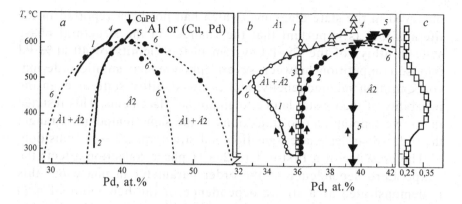

Fig. 4.14. Section of the state diagram of alloys of the Cu–Pd system [55] (*a*); dependences on the concentration in Cu–Pd alloys near the composition of 40 at.% Pd in phases *B*2 and *A*1 (*b*), calculated from the temperature dependences of the lattice parameters of phases *A*1 (curves 1 and 4) and *B*2 (curves 2 and 5); curve 3 is obtained according to the 'lever' rule from curves 1 and 2 and characterizes the concentration composition in the alloy; curve 6 – according to [55]. The half-width of the Bragg reflection (111) of phase *A*1 (*c*) upon heating of the quenched Cu alloy is 36 at.% Pd.

It was determined that in the coexisting *A*1 and *B*2 phases in the two-phase region, there is a small periodic (n.b.) concentration redistribution of the alloy components, which is manifested in the appearance of satellites near the main lines, but does not cause any noticeable broadening of the lines themselves.

The appearance of the satellites near structural lines corresponds to processes associated with low stability or instability of the crystal lattice relative to concentration waves [51]. Calculation within the framework of a simple model of one-dimensional sinusoidal modulations gives very different values of the concentration fluctuation wavelengths (modulation period) in the *B*2 ($\lambda \sim (24 \pm 3)$ nm) and *A*1 ($\lambda \sim (7\pm3)$ nm) phases. The maximum amplitude of atomic displacements caused by the periodic distribution of alloy components in phases is ~0.03 nm, which is close to the values of the total r.m.s. atomic displacements.

It is important to note that the general regularity observed in the two-phase region is the presence of concentration inhomogeneities in the studied alloys during the $B2{\rightarrow}A1$ and $A1_{quench}{\rightarrow}B2{\rightarrow}A1$ phase transition, which have their own characteristics.

Thus, when studying the structure-phase states of Cu–Pd alloys in the region of 40 at.% Pd, extraordinary features were observed in the state diagrams, which had not been reported in the literature. It is important that the changes in the structural phase state in alloys of the Cu–Pd system in the vicinity of 40 at.% Pd impose competition between several phenomena: atomic ordering, concentration inhomogeneity, and a low-stability state or even the instability of the crystal lattice to atomic displacements. This resu;ted in an overlapping of two transitions in the same temperature range: the order–disorder phase transition and structural $B2 \rightarrow A1$ transition.

The presence of a first-order $B2 \rightarrow A1$ phase transition affects the temperature dependence of the order parameter in phase $B2$; this is demonstrated as a strong dependence of the behaviour of $\eta(T)$ on the transition direction and alloy concentration. The $B2 \rightarrow A1$ phase transition always passes through a two-phase state and is accompanied by a concentration-induced separation of the $B2$ and $A1$ phases, and by the concentration heterogeneity. The nature of the concentration inhomogeneity depends on the initial structural phase state.

Anisotropy of atomic displacements. Using the R.W. James method [58], the r.m.s. displacements of atoms from the sites of the middle lattice along different crystallographic directions in the phase with the $A1$ structure (along <100>, <110> and <111>) were calculated from the experimental data on the temperature dependences of Bragg reflections, and the results of this calculation are presented in projections on the (110) plane (Fig. 4.15). It can be seen that with increasing temperature in the region of low-stability states preceding the irreversible (monotropic) $A1_{quench} \rightarrow B2$ phase transition (similarly, to the case after the reversible $B2 \leftrightarrow A1$ phase transition), the total r.m.s. atomic displacements increase and so does their anisotropy. In the range of temperatures 50°C higher than the end of the $B2–A1$ phase transition, the anisotropy of the r.m.s. displacements disappears, and the total r.m.s. displacements of the atoms increase.

The results of analysis of atomic displacements in the (110) cross sections of the FCC lattice taking into account their size and unit cell are interesting, which indicate a pronounced anisotropy of the total atomic displacements in low-stability pre-transitional regions.

The anisotropy and the value of the averaged atomic displacements increase sharply when approaching the onset temperature of the $A1_{quench} \rightarrow B2$ phase transition. Note that deviation from the linear dependence of the lattice parameter on temperature in a disordered solid solution occurs in the same temperature range. These phenomena

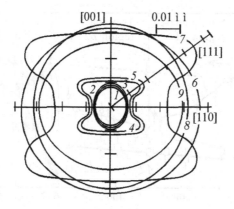

Fig. 4.15. RMS atomic displacements depending on various crystallographic directions in the (110) plane in phase $A1$ at various temperatures T, C: 1 – 50; 2 to 150; 3 to 200; 4 – 240; 5 – 260; 6 – 650; 7 to 700; 8 – 750; 9 – 800.

indicate that before the phase transition there is a decrease in the stability (a low-stability state appears) of the crystal lattice of the solid solution with respect to the displacement of atoms from the nodes of the crystal lattice.

From the temperature dependences of the Debye–Waller factor calculated from the main interference lines of the ordered $B2$ phase (without separating the displacements of atoms of different types along the sublattices in the $B2$ superstructure) and the disordered phase during the $A1_{quench} \rightarrow B2$ phase transition, the temperature dependences of the Debye temperature Θ_D were calculated (Fig. 4.16).

The temperature Θ_D (calculated in the isotropic approximation in the disordered $A1$ phase) before the $A1_{quenc} \rightarrow (B2 + A1)$ transition has a non-linear dependence (Fig. 4.16, curve 1). The observed decrease in Θ_D in phase $A1$ before the transition (to values close to the Debye temperature in phase $B2$) corresponds to a decrease in the stability (a low-stability state appears) of the crystal lattice. Thus, an analysis of the temperature dependence of the Debye temperature calculated in the isotropic approximation has shown that in the disordered phase there are low-stability pre-transitional regions.

Displacements of atoms in the ordered $B2$ phase in the two-phase region. In the ordered phase with the $B2$ structure formed by heating to temperatures above 300°C after the $A1_{quench} \rightarrow B2$ phase transition, the r.m.s. atomic displacements were measured in different sublattices occupied by palladium and copper atoms (Fig. 4.17).

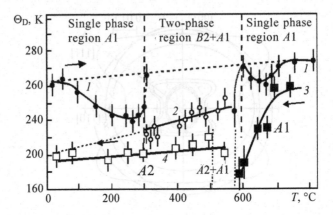

Fig. 4.16. Temperature dependences of the Debye temperature in phases *A*1 (1) and *B*2 (2) upon heating of the hardened Cu–36 at.% Pd alloy during the $A1_{\text{quench}} \rightarrow$ $(B2 + A1) \rightarrow B2$ transitions and phases *A*1 (3) and *B*2 (4) during cooling of the Cu–39.5 at.% Pd alloy with experiencing an $A1 \rightarrow B2$ phase transition.

The r.m.s displacements of copper atoms are larger than those of palladium, and the ratio $<u^2_{\text{tot}}>\text{Cu}/<u^2_{\text{tot}}>$ Pd = 1.9 is inversely proportional to the mass ratio of these atoms $(m_{\text{Cu}}/m_{\text{Pd}})^{-1} = 1.7$.

The r.m.s. atomic displacements linearly increase with increasing temperature in the range 300–500°C. The singularity is observed only in a narrow region immediately after the $A1_{\text{quench}} \rightarrow B2$ transition, where with increasing temperature a nonmonotonic change in the total r.m.s. displacements in different sublattices of the *B*2 superstructure takes place.

Manifestation of the effects of anharmonicity of atomic vibrations in the crystal lattice during phase transitions in CuPd alloys. The results of experimental studies to determine the temperature dependences of the integral characteristics (Debye–Waller factor, atomic displacements, and characteristic temperature) can serve as the basis for the analysis of interatomic interactions in crystals. It is known [59] that the nonlinearity of the temperature dependences of integral dynamic characteristics is due to the anharmonicity of integral interatomic interactions. An analysis of the dynamic parameters and the change in the structural states in CuPd alloys revealed a certain correlation. A change in the structural state of a phase is accompanied by a change in its dynamic properties.

In CuPd alloys whose structure mainly changes through a BCC↔FCC phase transition, the r.m.s. atomic displacements of atoms in the temperature range in the region of low-stability states

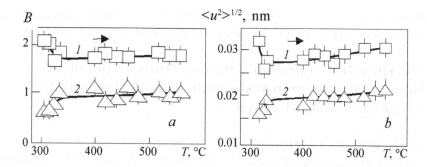

Fig. 4.17. Temperature dependences of (*a*) Debye–Waller factor and total r.m.s. atomic of atoms (*b*) in phase *B*2 after the *A*1$_{quench}$ →(*B*2 + *A*1) phase transition in the Cu–36 at.% Pd alloy (*a* – curves 1 and 2 are calculated from the intensities of the main and superstructural lines, respectively; *b* – curves 1 and 2 of the displacements of Cu and Pd atoms on the sublattices of *B*2 superstructure, respectively).

preceding the phase transition are anisotropic. The 'softening' of the lattice in the FCC phase is observed in the <111> direction, i.e., atoms in the structure of the FCC lattice prepare for rearrangement into a BCC coordination.

Thus, preparatory phenomena and low-stability states are observed in the high-temperature region, in the vicinity of the structural phase transition. A few tens of degrees before the temperature of the onset of the structural phase transition, intraphase transformations occur, which prepare the phase for the upcoming transition [59].

Conclusion. Low-stability states and structural changes in the region of phase transitions have been studied using Cu–Pd alloys as an example at a concentration of 40 at.% Pd. To this end, using the in situ X-ray diffraction patterns obtained directly in the temperature regions of the *B*2–*A*1 structural transition and the order–disorder phase transition we have determined – crystal lattice parameters, atomic long-range order parameters, and the Debye–Waller factor in the *B*2 and *A*1 phases. Based on the X-ray structural data, an analysis was made of the features of the *B*2 and *A*1 phases in CuPd alloys near structural phase transitions and the behaviour of atomic vibrations in the crystal lattice was discussed. It is has been shown that low-stability states are realized in the vicinity of structural-phase transformations in CuPd alloys in the region of 40 at.% Pd in which a number of anomalous phenomena are observed (for example, anisotropy of atomic displacements, concentration inhomogeneities, stratification, heterophase fluctuations, nonlinearities in the dependences of the lattice parameters and long-range order parameters, etc.), which prepares the system for transformation.

4.4. The effect of periodic stacking faults on the shear-induced mechanism transformations in the low-stability state of a condensed system

In addition to the influence of point defects and their complexes, here we discuss planar crystal lattice defects, in particular, stacking faults. For this case, the computational grid was selected in the form of a rectangular parallelepiped with sides oriented along the <110>, <$\bar{1}$10> and <001> directions of the BCC lattice. A stacking fault was formed by shifting a number of {110} planes of the BCC lattice in the <$\bar{1}$10> direction by a distance of $a\sqrt{2}/4$, where a is the constant of the initial BCC lattice.

The calculations were carried out using the original interatomic interaction potentials selected in the form of a polynomial

$$\varphi(r) = \begin{cases} \dfrac{A_n}{r^n} + A_0 + A_1 r + A_2 r^2 + A_3 r^3 + A_m r^m, & r < r_c, \\ 0, & r \geq r_c, \end{cases}$$

where r_c is the cutoff radius of the interaction potential, which was chosen between the third and fourth coordination spheres. The potential parameters were determined from the given values of the lattice constant, binding energy, and elastic moduli. The selected form of the interatomic interaction potential provides a decrease in the elastic constants C_{11} and C_{12}, as well as the shear modulus C' with a decrease in the bulk modulus B. In this case, there is an increase in the atomic displacements in the vicinity of the defect which corresponds to the delocalization of the static atomic displacements with a decrease in the elastic constants.

The potentials of interatomic interaction were used for two sets of values of their parameters, which are given in Table 4.3 for $m = n = 4$.

Specifically, a stable state of the BCC system was used in one case, and a low-stability pre-transitional state with abnormally low values of the elastic moduli in the other case. The calculations showed that for a large value of the bulk modulus of elasticity B, the introduced stacking fault does not cause any perturbations in the lattice sufficient to initiate the transition to a new phase. The initial BCC structure remains stable – the planes displaced during the process of relaxation again occupy their original positions. With a small value of the modulus B and a correspondingly low shear modulus C', the BCC lattice becomes unstable to the shift of the {110} planes in the <$\bar{1}$10> direction. A stacking fault causes a martensitic type transformation, carried out by shifting {110}-type planes in the <$\bar{1}$10> direction. The final structure of the newly formed phase depends on the distance between the defects. Figures 4.18–4.21 show the {110} plane displacement schemes for computational cells of various sizes and, therefore, different distances between stacking faults. It is evident that at a small distance between defects, the transformation occurs by twinning. Inside the twin, the sequence of planes is $ABCABC$ ..., which corresponds to the FCC structure. The {110} plane of the BCC lattice becomes a close-packed plane of the {111} face-centred cubic structure. The arrangement of atoms in two adjacent planes {110} of the initial structure after the transition is shown in Fig. 4.22 for the case of a computational cell of size 4×4×4 containing 256 atoms.

The displacement of the {110} planes relative to each other is accompanied by compression of the BCC lattice along the <001> direction by 9% and elongation along the <$\bar{1}$10> direction by ~11%. Such deformation is necessary for the transformation of the {110} plane of the BCC lattice into the {111} plane of the FCC lattice, i.e., for the formation of a hexagon from six nearest neighbours in one plane (Fig. 4.22). The formation of the remaining six nearest atomic neighbours (three in neighbouring planes) is accompanied by compression of the BCC lattice in the {110} direction by 9%.

Table 4.3. Potential parameters for $m = n = 4$

Potential parameters $\varphi(r)$	$B = 1.73 \cdot 10^{11}$ N/m^2	$B = 0.7 \cdot 10^{11}$ N/m^2
An	78.67722	−27.34020
A0	−3.78256	88.91930
A1	−6.38268	−103.66110
A2	5.19268	44.08128
A3	−1.21207	−8.12028
Am	0.09172	0.54857
rc	4.5856	4.5856

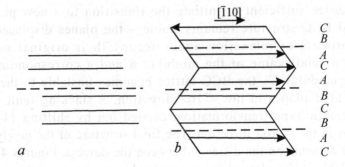

Fig. 4.18. Scheme of displacements of the {110} planes of the BCC lattice in the projection onto the {001} plane (the distance between stacking defects is $3d_{\{110\}}$): a – initial position of the planes, b – position after relaxation.

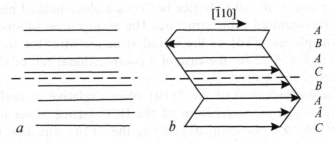

Fig. 4.19. Scheme of displacements of the {110} planes of the BCC lattice in the projection onto the {001} plane (the distance between stacking faults is $4d_{\{110\}}$): a – initial position of the planes, b – position after relaxation.

When the distance between the defects increases, there is no distinct twinning. The final structure is a non-repeating sequence of close-packed planes and contains fragments of both the FCC and HCP structures (Fig. 4.21).

In addition, the packing sequence of the planes varies depending on the distance between the defects. Similar structures have almost equal energies, implying that they are thermodynamically close, although they differ symmetrically. Their coordination numbers are not characteristic of HCP or FCC structures and can be equal 5 or 10, or even a non-integer (Fig. 4.23). The formation of such polytype structures in the FCC lattice is discussed in [60–61].

In addition to the displacement of the {110} planes, the transformation is accompanied by a change in volume, which was ~6% for the interatomic interaction potential used. This is due to a peculiar character of the dependence of the energy of the BCC and

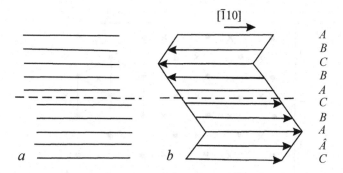

[Ī10]

	A
	B
	C
	B
	A
	C
	B
	A
	Â
	C

a   b

Fig. 4.20. Scheme of displacements of the {110} planes of a BCC lattice in projection onto the {001} plane (the distance between packing defects is $5d_{\{110\}}$): a – initial position of the planes, b - position after relaxation.

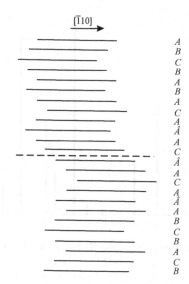

[Ī10]

A
B
C
B
A
B
A
C
A
Â
A
C
Â
A
C
A
Â
A
B
C
B
A
C
B

Fig. 4.21. Scheme of displacements of the {110} planes of a BCC lattice in projection to the {001} plane (the distance between packing defects is $12d_{\{110\}}$).

FCC structures on the atomic volume (Fig. 4.24). The FCC structure has a deeper minimum for a larger volume.

The calculations showed that for a fixed size of the simulated cell, there are no transformations. This can be interpreted as follow. Under constrained system conditions or during loading, the thermodynamic difference between different structural states of the system increases,

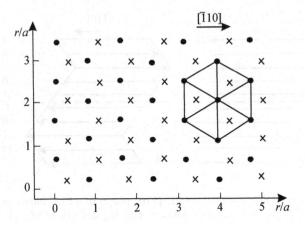

Fig. 4.22. Arrangement of atoms in two adjacent {110} BCC planes after the transition (circles and crosses indicate atoms of the planes *B* and *A*.

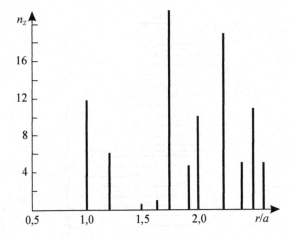

Fig. 4.23. Atomic distribution over the distance for the polytype structure shown in Fig. 4.20.

i.e., the probability of a transformation decreases. This might result in the fact that the transformation temperature would shift to the region of higher values, which is observed experimentally [62].

Conclusion. In a system with low elastic moduli, large static displacements of atoms from lattice sites are observed around defects. The resulting lengthy strain fields lead to the interaction of defects with each other. As a result of this interaction, it is possible to achieve ordering the defects by lowering the elastic energy of the system. It has been shown that defects preserving the symmetry of the initial BCC structure stabilize the initial structure. Defects that

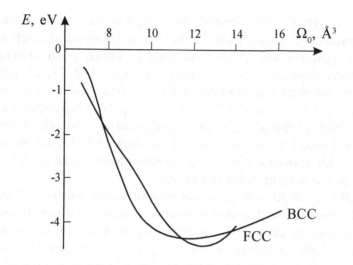

Fig. 4.24. Dependence of the energy E of the BCC and FCC structures on the atomic volume Ω_0 at a low value of the volume modulus $B = 0.7 \cdot 10^{11}$ N/m^2.

violate the symmetry of the initial structure can initiate a martensitic type transition in a lattice with low elastic moduli. It is the defecrt symmetry which ensures the transformation into a definite structure.

In the stable state of the condensed system (rigid lattice), the displacement fields in the vicinity of the defect are localized and no transformation is observed.

The molecular dynamics (MD) method is used to study the features of the effect of planar structural defects on the structural rearrangements of bcc alloys in a low-stability state.

It is shown that in a low-stability pre-transitional state of a BCC system, the interaction of packing defects can cause a transformation by a shear mechanism. At small distances between stacking defects, the transformation occurs by twinning. The possibility of the formation of polytype structures has been demonstrated.

4.5. The influence of stacking faults on the stability of *B2* alloys

In this section we analyze the effect of stacking faults on structural transformations in a binary BCC alloy. Martensitic transformations are observed in many alloys with a *B2* structure, among which TiNi and alloys based on it occupy a special place. With decreasing temperature, they exhibit sequences of *B2→B19′* or *B2→R→B19′*

martensitic transformations depending on the composition and thermomechanical treatment [1–8]. Pre-transitionalal phenomena in these systems are quite pronounced. There is an anomalous temperature dependence of the elastic constants [10, 11, 25, 63], the transverse acoustic phonon branch TA2 <ξξ0>experiences softening, and a dip occurs at $q \approx 2/3$ [3–5, 12, 14]. It is assumed that the condensation of the $q \approx 2/3$<110> mode causes a transformation of the second kind into an incommensurate phase, which then undergoes a first-order transition into the rhombohedral R-phase which is similar to the second-order transition.

By alloying TiNi with atoms of iron or cobalt (instead of nickel) or by deviating from the stoichiometric composition towards nickel it is possible to change the martensitic transformation sequence from $B2 \rightarrow B19'$ to $B2 \rightarrow R \rightarrow B19'$. In [1], it is suggested that this may be due to the formation of long-range order in the distribution of impurity atoms. Here we study the effect of planar defects periodically arranged in the $B2$ structure, caused by a deviation from stoichiometry and a violation of long-range order in the arrangement of atoms, on a possible martensitic transformation. The Ti–Ni system was chosen as a model system.

In this section we deal with the planar defects in the form of stacking faults. The stacking fault was created in exactly the same way as in the case of a one-component system by displacing one part of the crystallite relative to the other in the {110} plane in the $<\bar{1}10>$ direction. Figure 4.25 shows a characteristic form of the change in crystallite energy as a function of the displacement value d.

Curve 1 corresponds to the atomic interaction potential whose parameters are determined at the experimental value of the shear modulus $C' = 0.23 \cdot 10^{11}$ N/m², and curve 2 – at $C' = 0.50 \cdot 10^{11}$ N/m² [20, 64–66]. Both curves have local minima corresponding to the displacement of planes at a distance $a\sqrt{2}/4$. It is seen that, at a lower value of C', the shear resistance of the {110} planes is smaller, and the energy of the structure with a stacking fault slightly differs from the energy of a defect-free structure. Since the energy of the defect is low, such a system will have a tendency to form stacking defects. It is possible that the interaction of defects with each other will contribute to the formation of long-period structures.

To verify this assumption, stacking faults in computational cells of different sizes were examined. The computational cell, whose edges are oriented along the x, y, z axes (Fig. 4.26 a), is a system of {110} planes (Fig. 4.26 b). The stacking fault is introduced by the

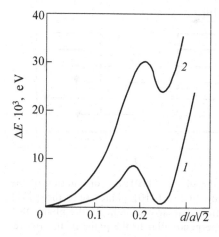

Fig. 4.25. Crystal energy change depending on the amount of shift d of the {110} planes.

displacement of half of the {110} crystallite planes relative to the other half by $a\sqrt{2}/4$. In a crystallite with a size of $4a\sqrt{2} \times 4a\sqrt{2} \times 4a$ and 256 atoms, a long-period structure was formed as a result of relaxation. This structure is formed by displacements of the {110} planes in the $<\bar{1}10>$ and $<001>$ directions. The scheme of displacements is shown in Fig. 4.26 b. In the Oy direction, the resulting structure represents alternating planes of the like atoms (Fig. 4.27 a). The arrangement of atoms in the {110} plane is shown in Fig. 4.27 b. Each atom is surrounded by six nearest neighbours, two of which are of the same sort, and four are of the other. The base of the unit cell is a rhombus with a side equal to twice the distance between the nearest neighbours.

The resulting structure corresponds to the global energy minimum. There are two local energy minima. The structure corresponding to the first minimum is formed by shuffling the {110} planes in the $<\bar{1}10>$ direction. Moreover, the displacements of Ni and Ti atoms are somewhat different. Nickel atoms experience the greatest displacements. Their displacements account for about 15.7% of the lattice parameter, while Ti atoms for ~11.6%. The projections of two adjacent {110} planes displaced relative to each other are shown in Fig. 4.28 a. Figure 4.28 b shows the projections of the {110} planes on the xy plane. The arrows indicate the direction of the plane shift, and the dotted line (Fig. 4.28) indicates and the dotted line shows the initially introduced stacking fault.

The second local minimum corresponds to a metastab+le structure formed in addition to shuffling displacements of individual {110} planes in the $<\bar{1}10>$ direction by additional atomic displacements in

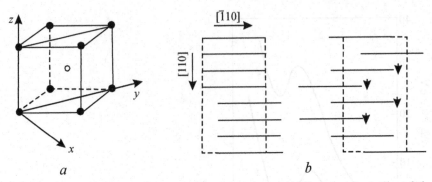

Fig. 4.26. Scheme of the displacement of the planes {110}: *a* – orientation of the axes of the simulated block; *b* – projection of the {110} planes onto the {001} plane before (left) and after (right) relaxation. The arrows indicate the planes that underwent a shift in the direction <001>, perpendicular to the plane of the figure.

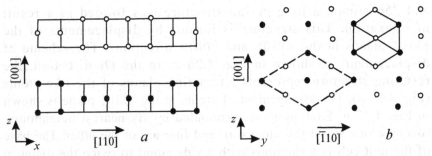

Fig. 4.27. Scheme of alternating planes of the like atoms: *a* – planes alternating in the {$\bar{1}\bar{1}0$} direction; *b* – the arrangement of atoms in one of the {110} planes.

the <110> direction (Fig. 4.29). The displacements of nickel atoms are 2 times larger than those of Ti atoms. The shift of Ni atoms in the <$\bar{1}10$> direction is ~17.5%, and in the <110> direction ~7.8% of the lattice constant.

Conclusion. An introduction of a stacking fault giving rise to displacements of one part of the crystallite relative to the other in the {110} plane in the <$\bar{1}10$>direction does not cause significant changes in the energy of the structure with stacking faults from that of a defect-free structure.

However, the resulting structure is a structure with a global energy minimum. In addition to the minimum, there are two local energy minima. The structure corresponding to the first minimum is formed by shuffling the {110} planes in the <$\bar{1}10$>direction.

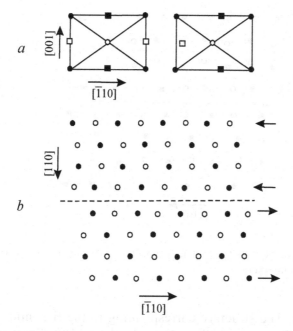

Fig. 4.28. Displacements of the {110} planes of the B2 structure: *a* – arrangement of atoms in two adjacent {110} planes before (left) and after (right) the shift; *b* — projections of the {110} planes onto the {001} plane. Ni atoms are located half a translation period lower than Ti atoms. Arrows indicate the direction of movement of the {110} planes; • and ■ are Ti atoms; ○ and □ - Ni.

Moreover, the displacements of Ni and Ti atoms are somewhat different. Nickel atoms experience the largest displacements. Their displacements account for about 15.7% of the lattice parameter, while Ti atoms for ~11.6%. The second local minimum corresponds to a metastable structure formed in addition to the shuffling displacements of individual {110} planes in the <$\bar{1}$10> direction by additional atomic displacements in the <110> direction. The displacements of the nickel atoms are 2 times larger than those of the Ti atoms. The shift of the Ni atoms in the <$\bar{1}$10> direction is ~17.5%, and in the <110> direction ~7.8% of the lattice constant.

It is has been shown in this part that the introduction of stacking faults giving rise displacements of one part of the crystallite relative to the other in the {110} plane in the <$\bar{1}$10> direction leads to the thermodynamic advantage of the structure containing a stacking fault, whose energy slightly differs from the energy of a defect-free structure. The resulting structure corresponds to the global minimum of energy. In addition to the global minimum, there are two local

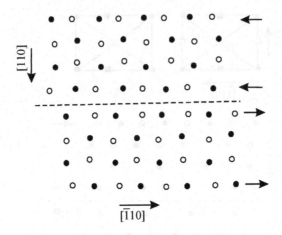

Fig. 4.29. Projections of the {110} planes onto the {001} plane, corresponding to the second local minimum.

energy minima. The structure corresponding to the first minimum is formed by shuffling the {110} planes in the <$\bar{1}$10> direction. The second local minimum corresponds to a metastable structure formed in addition to the shuffling displacements of individual {110} planes in the <$\bar{1}$10> direction by additional atomic displacements in the <110> direction.

4.6. The inheritance of the structural point defects by the $B2$ ω-type martensitic phase

The BCC alloys, among which a special place is occupied by TiNi and alloys based on present both theoretical and practical interest. With decreasing temperature, they exhibit experience of the $B2{\rightarrow}B19'$ or $B2{\rightarrow}R{\rightarrow}B19'$ martensitic transformations depending on the composition and thermomechanical treatment [1–6, 12]. Pre-transitional phenomena in the low-stability states of these systems are pronounced.

In this section, an attempt is made to study the effect of defects periodically arranged in the $B2$ structure, caused by a deviation from stoichiometry and a violation of long-range order in the arrangement of atoms, on a possible martensitic transformation. The Ti–Ni system was chosen as a model system. The effects of the inheritance of point defects of the $B2$ structure by the daughter phase during the ω-like martensitic transformation are studied, with particular attention being paid to the fine structure of the system in the presence of point defects and their complexes.

The interest in the issues of the inheritance of defects of the high-temperature phase is by the martensitic phase is due to the fact that they significantly affect the thermodynamic characteristics of the phases and the kinetics of transformation [67, 68]. There is an extensive class of materials, for example, such ordered intermetallic alloys with the $B2$ structure as NiTi, NiAl, FeMn, FeNi, CuNi, CuMn, AuCd, InTl, etc., in which the elementary plasticity act is carried out due to the reversible martensitic transformation and a number of other processes. For instance, these alloys can exhibit complete or partial reversibility of inelastic deformation [67, 68]. The recovery is initiated by the forces of mechanical and chemical origin (forces from stacking faults, and antiphase boundaries, and point defects).

While the mechanical stress is always directed toward the recovery of deformation, then the thermodynamic forces proportional to the difference in the thermodynamic potentials of the phases can be directed both ways depending on the temperature. Then the final pathway of the reaction or the direction of the deformation development will be determined by the sum of the mechanical and chemical stresses. It is important to note that chemical forces often prevail over mechanical ones, therefore microstresses, like stresses caused by external loads, have only a small disturbing effect on the development of the transformation. The appearance of defects can change the ratio of chemical and mechanical forces and enhance the effect of mechanical stresses on the transformation.

The phase transformation of a material containing defects causes the reaction product to accumulate high-energy defects as a result of their inheritance. In the 'exactly backward' reaction, a complete restoration of the initial structure with low energy occurs. Other orientational variants of the reverse transformation pathway result in a situation where the high-energy defects inherited in the direct reaction acquire even higher energy.

Consequently, the energy reasons appear in the lattice for the reverse martensitic transformation with a certain orientation of the transformation path, since only in such reactions the complete restoration of the initial defective structure with low energy occurs.

This conclusion turns out to be true not only for dislocations, but also for point defects or their complexes, when a change in the short-range order is significant [67].

The point defects can affect not only the path of the reverse martensitic transition, they can also change the sequence of direct martensitic transformation. For instance, TiNi containing 50.2 at.% Ni, experiences a $B2(\beta) \to B19'$, while in annealed alloys with an excess of nickel, for example $Ti_{49}Ni_{51}$, the martensitic transition occurs in the $B2 \to R \to B19'$ sequence [1, 3, 12] This is due to a strong influence of the regions of local enrichment with nickel atoms and a more rapid increase in the fraction of 'disordered' atoms in the disordered subsystem as compared with concentration-induced disordering. An increase in the number of local regions with defects upon annealing affects the sequence of transformation.

There is no theory explaining the laws of inheritance of defects. Issues of inheritance of dislocations during martensitic transformations were covered in [67, 68].

The interest in the study of point defects during martensitic transformations is due to the fact that in alloys with an ordered $B2$ (CsCl) structure, the degree of long-range order η is usually less than unity even for stoichiometric composition. For example, in TiNi-, the degree of long-range order η does not exceed 0.84 [12]. In these alloys, there is a significant number of point defects are observed associated with local disordering. Deviation from stoichiometry also lowers the degree of long-range order. Excess atoms, occupying a site in a foreign sublattice, form a substitution defect. When they are inherited by the martensite, not only the immediate environment around the defect changes, but, as shown in [20], the number of point defects might also increase.

The martensitic transformation can change the nearest neighbourhood around a point defect. As a result, if before the conversion the contribution from point defects to the thermodynamic potential was G_0, then after the transformation it changes to ΔG due to the difference in the energy of defects and their amount in the initial phase and in the reaction product. A change in ΔG leads to the appearance of additional chemical forces generated as a result of the martensitic transformation. In a pre-transitional low-stability state under conditions of a 'soft' lattice (low shear moduli), the defects tend to form a superlattice of defects due to a decrease in the relaxation energy. The formation of such a superlattice in a pre-transitional low-stability state can stimulate the martensitic transformation.

To study the inheritance of point defects, we used the molecular dynamics method [66, 69, 70] with a computational cell with periodic boundary conditions, which that changes its size and shape. In the pre-transitional low-stability state under conditions of a soft lattice, distortions around the defect reach the boundaries of the computational cell, and the defects cannot be any longer considered as isolated defects. In this case, the use of periodic boundary conditions is equivalent to considering an infinite superlattice of defects with primitive translation vectors equal to the edges of the computational cell. Modelling of the processes of martensitic transformation in a crystal imposes some limitations associated with periodic boundary conditions caused by the synchronism of processes in all primitive cells of the superlattice of defects.

The study has been carried out with point defects in the $B2$ structure, whose introduction caused a martensitic transition to a hexagonal layered ordered structure.

The following types of defects were investigated:

– an order defect (Fig. 4.30 a) formed by rearrangement of Ti and Ni atoms on the first coordination sphere;

– an anti-structural Ti atom and an order defect forming a linear chain along the <111> direction (Fig. 4.30 b);

Schematically the displacements of the planes of the $B2$ lattice upon transition to the martensitic phase are shown in Fig. 4.31 a, b showing the stacking of the $\{111\}_{B2}$ planes in the computational cell. The planes occupied by Ni are indicated by solid circles, Ti – open circles. The position of the planes containing defects and the type of atom forming the defect are indicated by additional circles with numbers. The arrows indicate the direction and magnitude of ω-like displacements of the close-packed planes of the $B2$ lattice and the atoms of the defect during the martensitic transition. Figure 4.31 c, d presents the diagram of the planes of the formed phase. The circlesin the bottom part characterize the type of plane, and the circles in the top part indicate the location of atoms on the $<111>_{B2}$ line containing defects.

The numbers indicate the positions of the atoms that make up the defect before and after the transition.

From Fig. 4.31 a it can be seen that the atoms with numbers 2 and 3, which form order defects in the $B2$ lattice, after its transformation to martensite, lie in the planes of their type and are no longer defects (in the traditional sense) in the new phase, but there is a linear chain of the order defects made up of four atoms along the C axis of the hexagonal phase. The number of atoms forming the defect has doubled. This lowered the degree of order

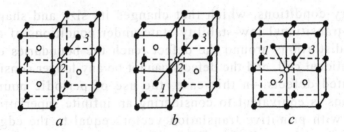

Fig. 4.30. Structure of point defects in a *B*2 lattice is a defect forming an isosceles triangle with base *a*, lying in the {110} plane, whose vertices are formed by an excess Ti atom and an order defect.

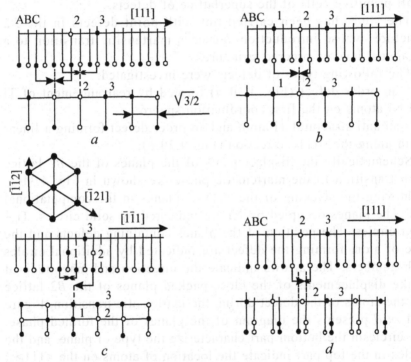

Fig. 4.31. Scheme of displacements of planes in an ω-like transformation.

in the martensitic phase and increased the chemical component of the energy of the system associated with defects. For the defect presented in Fig. 4.30 *b*, as follows from Fig. 4.31 *b*, the number of atoms forming the defect has not changed; only a shift in its position occurred. In the martensitic transformation of the lattice with the defect shown in Fig. 4.30 *c*, after the transition, the excess Ni atom remains an anti-structural atom (Fig. 4.31 *c*). On the $\{111\}_{B2}$ line passing through atoms 2 and 4, a linear chain of four atoms was formed representing an order defect. The number of defects has changed. Of the three 'disordered' atoms, five 'defective' atoms were formed: an excess atom and a linear chain of four atoms relatively remote from it. Figure 4.31 d is a diagram of the transition of the $B2$ phase into a ω-like ordered structure presented for the interaction potential adjusted to a more rigid value of the shear modulus C'. Shear modulus has been increased 4 times. The letters above the planes indicate the type of solid-sphere stacking. The martensitic structure is obtained by the collapse of the planes A and C with the formation of a plane of the like atoms, higher than in the neighboring one. The atoms of the plane closest to it are projected into the centre of the hexagon. The numbers indicate the position of atoms that representing the former defects in the $B2$ phase. As can be seen from Fig. 4.31 d, a linear chain of atoms has been formed which are displaced from their planes. The number of 'defective' atoms has increased threefold.

Conclusion. It has been shown that the structure of a point defect inherited by martensite changes depending on the final reaction product. The number of point defects of the order during martensitic transformation, as a rule, increases. High-energy defects are formed from linear chains oriented along the <111>-type direction. These defects contribute to the fact that the reverse martensitic transformation occurs exactly backward.

A change in the number of defects during the martensitic transition leads to the accumulation of the energy associated with the chemical contribution to the thermodynamic potential by the defects, since the number of pairs of atoms of the same type in the first coordination spheres changes. The displacements of atoms around defects in the martensite are more localized than in the high-temperature phase. Therefore, the elastic energy associated with the non-chemical contribution from the defect decreases in the course of the $B2 \rightarrow \omega$-like martensite transition.

Thus, for order defects, anti-structural atoms, and their simplest complexes, it has been shown that in some cases the number of point defects increases during martensitic transformation. At the same time, the proportion of chemical energy accumulated by defects increases. Some types of defects in the $B2$ structure in martensite correspond to the appearance of high-energy linear disordered chains in the типа type direction. This can play a significant role in determining the path of reverse martensitic transformation.

When studying the inheritance of the structural point defects of the $B2$-type by the ω-type martensitic phase, it has been found that the structure of the point defect inherited by the martensite changes depending on the final reaction product. The number of point order defects during martensitic transformation commonly increases. High-energy defects are formed from linear chains oriented along the <111>-type direction. These defects give rise to an exactly backward path of the reverse martensitic transformation.

A change in the number of defects during the martensitic transition leads to the accumulation of the energy, associated with the chemical contribution to the thermodynamic potential, by the defects, since the number of pairs of atoms of the same type in the first coordination spheres changes. The displacements of atoms around defects in the martensite are more localized than in the high-temperature phase. Therefore, the elastic energy associated with the non-chemical contribution from the defect decreases during the $B2 \rightarrow \omega$-like martensite transition.

Summary

Using computer simulation methods, it has been shown that in the low-stability state of a condensed system, the cooperative interaction of point defects can lead to their ordered arrangement, and the resulting static displacement fields can both stabilize the $B2$ structure and contribute to its instability and martensitic phase transition. In the presence of a certain type of defects in the $B2$ structure, the latter is unstable to shifts of the $\{111\}$ planes along the direction of the <111> type. In the final structure, the displacement fields around the defects are localized, and the defects themselves organically fit into the structure of the resulting phase. It should be noted that structural defects of the parent phase become natural elements of the structure of the final daughter phase.

When studying the possibilities of the formation of a columnar structure of a relaxation type in BCC alloys, it was shown that the formation of a low-stability equilibrium columnar antiphase structure of a relaxation type in BCC systems is energetically quite possible. A necessary condition for this structure to form in the alloys with a basic $B2$ superstructure is the presence of anisotropy (e.g., interatomic interaction). Moreover, a whole set of columnar structures of various sizes provides the energy advantage of a low-stability long-period state of relaxation type. It is very likely that under a small thermal-force action the system will undergo structure-phase transformations, and at non-zero temperatures, a certain set of low-stability long-period states will coexist.

We have studied low-stability pre-transitional states, order–disorder phase transitions and $B2$–$A1$ structural transformations in Cu–40 at.% Pd alloys as well as some peculiar features of structural changes in the region of phase transitions directly in the temperature regions of the $B2$–$A1$ structural transition and order–disorder phase transition. On the basis of X-ray structural data, an analysis has been made of the features of the $B2$ and $A1$ phases in the CuPd alloys near structural phase transitions and the behaviour of atomic vibrations in the crystal lattice has been discussed. It is shown that in the vicinity of structural-phase transformations in the CuPd alloys in the region of 40 at.% Pd low-stability states are formed in which a number of anomalous phenomena are observed (for example, anisotropy of atomic displacements, concentration inhomogeneities, stratification, heterophase fluctuations, nonlinearities in the dependences of the lattice parameters and long-range order parameters, etc.) which prepare the system for transformation.

The molecular dynamics method has been used to study the features of the influence of planar structural defects (stacking faults) on structural rearrangements in the BCC alloys in a low-stability state resulting in a transformation by the shear mechanism.

At small distances between the stacking faults, the transformation occurs by twinning. The possibility of the formation of polytype structures has been demonstrated.

Using $B2$-alloys as examples, the relationship between the presence of planar structural defects (stacking faults) and the stability of alloys relative to structural-phase transformations in a low-stability state of systems has been studied. It has been shown that during martensitic type transformations in BCC systems, the appearance of stacking faults introducing displacements of one part of the crystallite relative to the other in the {110} plane in the <110> direction leads to the thermodynamic advantage of the stacking fault structure, which slightly differs from the energy of a defect-free structure. The resulting structure corresponds to the global energy minimum. There are two local energy minima. The structure corresponding to the first minimum is formed by shuffling the {110} planes in the <$\bar{1}$10> direction. The second local minimum corresponds to a metastable structure formed in addition to the shuffling displacements of the individual {110} planes in the <$\bar{1}$10> direction by additional atomic displacements in the <110> direction.

Using the example of point and planar defects and their complexes, the inheritance of structural defects by the daughter phase during structure-phase transformations in the pre-transitional low-stability state of metal BCC systems has been studied.

During martensitic transformations in the presence of order defects, anti-structural atoms, and their simplest complexes, it has been shown that in a number of cases during martensitic transformations the number of point defects increases. It should be noted that the fraction of chemical energy accumulated by the defects increases. The structure of a point defect inherited by the martensite varies depending on the final reaction product. Some types of defects in the $B2$ structure in the martensite correspond to the appearance of high-energy linear disordered chains in the <111>-type direction. This can play a significant role in determining the path of the reverse martensitic transformation.

For order defects, anti-structural atoms, and their simplest complexes, it has been shown that in some cases the number of point defects increases during martensitic transformation. At the same time, the proportion of chemical energy accumulated by the defects increases. Some types of defects in the $B2$ structure in the martensite are accompaied by the formation of high-energy linear disordered chains in the <111> type direction. This can play a significant role in determining the path of reverse martensitic transformation.

The structure of a point defect inherited by the martensite varies depending on the final reaction product. The number of point order defects during martensitic transformation generally increases. High-energy defects are formed from linear chains oriented along the <111>-type direction. These defects cause the reverse martensitic transformation in the exactly backwards pathway.

A change in the number of defects during the martensitic transition leads to the accumulation of the energy associated with the chemical contribution to the thermodynamic potential, since the number of pairs of atoms of the same type in the first coordination spheres changes. The atomic displacements of atoms around the defects in the martensite are more localized than in the high-temperature phase. Therefore, the elastic energy associated with the non-chemical contribution from the defect decreases during the $B2 \to \omega$-like martensite transition.

When studying the possibilities of the formation of a columnar structure of a relaxation type in bcc alloys, it was shown that the formation of a low-stability equilibrium columnar antiphase structure of a relaxation type in bcc systems is energetically quite possible. A necessary condition for its implementation in alloys with a basic $B2$

superstructure is the presence of anisotropy (for example, interatomic interaction). Moreover, a whole set of columnar structures of various sizes provides the energy advantage of a low-stability long-period state of relaxation type. It can be expected that under a small thermo-force action the system will undergo structure-phase transformations, and at non-zero temperatures, a certain set of low-stability long-period states will coexist.

We studied low-stability pre-transitional states, order–disorder phase transitions and B2–A1 structural transformations in Cu–40 at.% Pd alloys, and features of structural changes in the region of phase transitions directly in the temperature regions of the $B2$–$A1$ structural transition and order–disorder phase transition. On the basis of X-ray structural data, an analysis was made of the features of the $B2$ and $A1$ phases in CuPd alloys near structural phase transitions and a picture of the behaviour of atomic vibrations in the crystal lattice was considered. It is shown that in the vicinity of structure-phase transformations in CuPd alloys in the region of 40 at.% Pd low-stability states are realized in which a whole spectrum of anomalous phenomena is observed (for example, anisotropy of atomic displacements, concentration inhomogeneities, stratification, heterophase fluctuations, nonlinearities in the dependences of the lattice parameters and long-range order parameters, etc.) that prepare the system for transformation.

The molecular dynamics method is used to study the features of the influence of planar structural defects (stacking faults) on structural rearrangements by the shear mechanism of bcc alloys in a low-stability state.

It is shown that in a low-stability pre-transitional state of a bcc system, the interaction of packing defects can cause a transformation by a shear mechanism. At small distances between packing defects, the transformation occurs by twinning. The possibility of the formation of polytype structures is demonstrated.

Using the $B2$-alloys as examples, the relationship between the presence of planar structural defects (stacking faults) and the stability of alloys relative to structure-phase transformations in a low-stability state of systems is studied. It is shown that during martensitic type transformations in BCC systems, the appearance of stacking faults introducing shifts of one part of the crystallite relative to the other in the {110} plane in the <110> direction leads to the thermodynamic profitability of the stacking fault structure, which slightly differs from the energy defect-free structure. The resulting structure corresponds to the global energy minimum. As it is achieved, there are two local energy minima. The structure corresponding to the first minimum is formed by shuffling the {110} planes in the <110> direction. The second local minimum corresponds to a metastable structure formed in addition to the shuffling displacements of individual {110} planes in the <110> direction by additional atomic displacements in the <110> direction.

Using the example of point and planar defects and their complexes, the inheritance of structural defects by the daughter phase during structure-phase transformations in the pre-transitional low-stability state of BCC metal systems has been studied.

During martensitic transformations for order defects, anti-structural atoms, and their simplest complexes, it has been shown that in a number of cases during martensitic transformations the number of point defects increases. At the same time, the proportion of chemical energy accumulated by defects increases. The structure of a point defect inherited by martensite varies depending on the final reaction product. Some types of defects in the $B2$ structure in martensite correspond to the appearance of high-energy linear disordered chains in the <111>-type direction. This can play a significant role in determining the path of reverse martensitic transformation.

For order defects, anti-structural atoms, and their simplest complexes, it has been shown that in some cases the number of point defects increases during martensitic transformation. At the same time, the proportion of chemical energy accumulated by defects increases. Some types of defects in the $B2$ structure in martensite correspond to the appearance of high-energy linear disordered chains in the <111> type direction. This can play a significant role in determining the path of reverse martensitic transformation.

The structure of a point defect inherited by martensite varies depending on the final reaction product. The number of point defects of the order during the martensitic transformation, as a rule, increases. High-energy defects are formed from linear chains oriented along the <111>-type direction. These defects contribute to the fact that the reverse martensitic transformation occurs exactly backward.

A change in the number of defects during the martensitic transition leads to the accumulation by the defects of energy associated with the chemical contribution to the thermodynamic potential, since the number of pairs of atoms of the same type in the first coordination spheres changes. The displacements of atoms around defects in martensite are more localized than in the high-temperature phase.

Therefore, the elastic energy associated with the non-chemical contribution from the defect decreases with the transition of the $B2$ ω-like martensite.

References

1. Lotkov A.I., Grishkov V.N. // Fiz. Met. Metalloved. - 1990. - No. 7. - P. 88–94.
2. Li D.Y., Wu X.F., Ko T. // Phil. Mag. A. - 1991. - V. 63. - No. 3. - P. 603-616.
3. Lotkov A.I., Grishkov V.N. // Izv. Univ. Fizika. - 1985. - V. 28. - No. 5. - P. 68–87.
4. Grishkov V.N., Lotkov A.I. // Fiz. Met. Metalloved. - 1985. - V. 60. - No. 2. - P. 351–355.
5. Lotkov A.I., Grishkov V.N. // Izv. Univ. Fizika. - 1991. - V. 34. - No. 2. - P. 106–121.
6. Lotkov A.I., Kuznetsov A.V. // Fiz. Met. Metalloved. - 1988. - V. 66. - No. 5. - P. 903–909.
7. Mercier O., Melton N., Greemafud G., Hagi J. // J. Appl. Phys. - 1980. - V. 51. - No. 3. - P. 1833–1834.
8. Brill, T. M., Mittelbach S., Assmus W., et al. // J. Phys .: Condens. Matter. - 1991. - V. 3. - P. 9621–9627.
9. Khachin V.N., Muslov S.A., Pushin V.G., Kondratiev V.V. // Metallofizika. - 1988. - V. 10. - No. 1. - P. 102–104.
10. Moine P., Allain J., Renker B. // J. Phys. F .: Metal Phys. - 1984. - V. 14. - P. 2517–2523.
11. Tietze H., Mullner M., Renker B. // J. Phys. C .: Solid State Phys. - 1984. - V. 17. - P. L529 – L532.
12. Dubinin S.F., Teploukhov S.G., Lotkov A.I. et al. // Fiz. Met. Metalloved - 1992. - No. 4. - P. 119–124.
13. Satija S.K., Shapiro S.M., Salamon M.B., Wayman C.M. // Phys. Rev.B. - 1984. - V. 29. - No. 11. - P. 6031-6035.
14. Ohba T., Shapiro S. M., Aoki S., Otsuka K. // J. Appl. Phys. - 1994. - V. 33. - P. L1631 – L1633.
15. Maeda K., Vitek V., Sutton A.P. // Acta Met. - 1982. - V. 30. - No. 11. - P. 2001 –2010.
16. Vitek V. // Phil. Mag. A. - 1988. - V. 58. - No. 1. - P. 193–212.
17. Yakovenkova L.I., Karkina L.E. Interaction of vacancies with a split dislocation in fcc metals // Theory and computer simulation of defective structures in crystals. - Sverdlovsk: UC AN SSSR, 1986. - 164 p.
18. Karkina L.E. // Fiz. Met. Metalloved. - 1989. - V. 68. - No. 3. - P. 459–465.
19. Zhorovkov M.F., Kulagina V.V. // Izv. Univ. Fizika. - 1992. - V. 35. - No. 1. - P. 3–8.
20. Zhorovkov M.F., Kulagina V.V. // Metally. - 1995. - No. 5. - P. 85–89.
21. Potekaev A.I., Naumov I.I., Kulagina V.V. et al., Natural long-period nanostructures / ed. A.I. Potekaev. - Tomsk: NTL Publishing House, 2002 .-- 260 p.
22. Potekaev A.I., Klopotov A.A. Kozlov E.V. et al., Low-stability pre-transitional structures in titanium nickelide. - Tomsk: NTL Publishing House, 2004 .-- 296 p.
23. Potekaev A.I., Dmitriev S.V., Kulagina V.V. et al., Low-stability long-period structures in metallic systems / ed. A.I. Potekaev. - Tomsk: NTL Publishing House, 2010 .-- 320 p.
24. Potekaev A.I., Starostenkov M.D., Glezer A.M. et al. Structural-phase states and properties of metallic systems / ed. A.I. Potekaeva. - Tomsk: NTL Publishing House, 2004 .-- 356 p.

25. Potekaev A.I., Starenchenko V.A., Kulagina V.V. and other Weak-resistant states of metal systems / ed. A.I. Potekaev. - Tomsk: NTL Publishing House, 2012 .-- 272 p.

26. Potekaev A.I., Starenchenko V.A., Kulagina V.V. and other Weak-resistant states of metal systems / ed. A.I. Potekaev. - Tomsk: NTL Publishing House, 2012.– 272 p.

27. Starenchenko S.V., Kozlov E.V., Starenchenko V.A. The laws of the thermal phase transition are order-disorder in alloys with superstructures L12, L12 (M), L12 (MM), D1a. - Tomsk: NTL Publishing House, 2007 .-- 268 p.

28. Matysina Z.A., Zaginaichenko S.Yu. Crystal structure defects: a monograph. - Dnepropetrovsk: Science and Education, 2003. - 284 p.

29. Potekaev A.I. // Izv. Univ. Fizika. - 1995. - V. 38. - No. 6. - P. 3–12.

30. Potekaev A.I. // Izv. Univ. Fizika. - 1996. - V. 39. - No. 6. - P. 22–40.

31. Potekaev A.I., Dudnik E.A., Starostenkov M.D., Popova L.A. // Izv. Univ. Fizika. - 2008. - V. 51. - No. 10. - P. 53–62.

32. Potekaev A.I., Kulagina V.V. // Izv. Univ. Fizika. - 2008. - V. 51. - No. 11/3. - P. 148-150.

33. Potekaev A.I. // Phys. Stat. Sol. (a). - 1992. - V. 134. - P. 317–334.

34. Kozlov E.V. // Izv. Univ. Fizika. - 1976. - V. 19. - No. 8. - P. 82–92.

35. Potekaev A.I., Naumov I.I., Kulagina V.V., et al. Low-stability metallic-based nano-structures // ed. A.I. Potekaev. - Tomsk: Scientific Technology Publishing House, 2018 .-- 236 p.

36. Potekaev A.I., Kulagina V.V. // Izv. universities. Physics. - 2009. - T. 52. - No. 8/2. - S. 456–459.

37. Klopotov A.A., Potekaev A.I., Gunter V.E., Kulagina V.V. // Izv. universities. Ferrous metallurgy. - 2010. - No. 10. - S. 61–67.

38. Potekaev A.I., Egorushkin V.E., Golosov N.S. // Phys. Stat. Sol. (a). - 1986. - V. 98. - P. 345–349.

39. Naumov II, Olemskoi AI, Potekaev A.I. // FMM. - 1993. - T. 75. - Vol. 6. - S. 47–57.

40. Lobodyuk V.A., Koval Yu.N., Pushin V.V. // FMM. - 2011. - T. 111. - No. 2. - S. 169–194.

41. Braddin D., Tendeloo Van G., et al. // Phil. Mag. A. - 1086. - V. 54. - No. 3. - R. 395-419.

42. Braddin D., Tendeloo Van G., Landuyt Van J., Amelinskx S. // Phil. Mag. B. - 1988. - V. 57. - No. 1. - P. 31–48.

43. Ogawa S. // Order-disorder transformations in alloys / eds. H. Waretimont, B. Heidelberg. - N.Y .: Springer, 1974. - P. 240–264.

44. Schastlivtsev V.M., Kaletina Yu.V., Fokina E.A. Martensitic transformation in a magnetic field. - Yekaterinburg: URORAN. - 2007. - 322 p.

45. Pushin V.V., Kondratiev V.V., Khachin V.N. Pre-transitional phenomena and martensitic transformations. - Yekaterinburg: URORAN, 1998 .-- 368 p.

46. Volkov A.Yu., Kazantsev V.A., Kourov N.I. // Physics of metals and metal science. - 2008. - No. 4. - S. 355–366.

47. Antonova O.V., Volkov A.Yu. // Physics of metals and metal science. - 2009. - V. 108. - No. 4. - P. 395–405.

48. Volkov A.Yu. // Physics of metals and metal science. - 2006. - No. 5. - P. 571–577.

49. Volkov A.Yu., Volkova E.G. // Materials Science. - 2006. - No. 6. - P. 25–31.

50. Klopotov A.A., Tailashev A.S., Kozlov E.V. // Izv. Univ. Fizika. - 1988. - V. 31. - No. 6. - P. 67–72.

51. Klopotov A.A., Tailashev A.S., Potekaev A.I., Kozlov E.V. // Izv. Univ. Fizika. - 1999. - V. 42. - No. 7. - P. 55–59

52. Klopotova A.A., Tailasheva A.S., Potekaev A.I. et al. // Izv. Univ. Fizika. - 1997. - V. 40. - No. 3. - P. 93–102.

53. Klopotov A.A., Tailashev A.S., Popov S.N., Kozlov E.V. // Izv. Univ. Fizika. - 1993. - V. 36. - No. 2. - P. 35–38

54. Volkov A.Yu., Antonova O.V,, Patselov A.M. // Deformation and destruction of materials. - 2007. - No. 4. - P. 20–26.

55. State diagrams of double metal systems / ed. N.P. Lyakishev. - M.: Mashinostroenie, 1996–2000. - V. 1–3.

56. Matveeva N.M., Kozlov E.V., Ordered phases in metallic systems. - M .: Science. - 1989 .-- 247 p.

57. Landau L.D., Lifshits E.M., Statistical Physics. - Moscow, Nauka, 1976. - Part 1. - 583 p.

58. James R. Optical Principles of X-Ray Diffraction. - M.: IL, 1950 .-- 572 p.

59. Tyapkin Yu.D., Lyasotsky I.V. Intrafase transformations // Itogi Nauki i Tekhniki. MITOM. - M.: VINITI AN USSR, 1981. - T. 15. - S. 47–110.

60. Potekaev A.I., Klopotov A.A., Kulagina V.V., Gunther V.E. // Izv. Univ. Ferr. Metallurgiya. - 2010. - No. 10. - P. 61–67.

61. Moody M., Ray J. R., Rahman A. // Phys. Rev. B: Condens. Matter. - 1987. - V. 35. - No. 2. - P. 557–570.

62. Potekaev A.I., Klopotov A.A., Matyunin A.N. and others // Deformation and destruction. - 2011. - No. 11. - P. 40–43.

63. Satija S.K., Shapiro S.M., Salamon M.B., Wayman C.M. // Phys. Rev. B. - 1984. - V. 29. - No. 11. - P. 6031-6035.

64. Kulagina V.V., Dudarev E.F. // Izv. Univ. Fizika. - 2000. - V. 43. - No. 6. - P. 58–63.

65. Kulagina V.V. // Izv. Univ. Fizika. - 2001. - V. 44. - No. 2. - P. 30–39.

66. Zhorovkov M.F., Kulagina V.V. // Izv. Univ. Fizika. - 1993. - V. 36. - No. 10. - P. 31–39.

67. Likhachev V.A. // Izv. Univ. Fizika. - 1985. - V. 28. - No. 5. - P. 21–40.

68. Likhachev V.A., Kuzmin S.A., Kamentseva Z.P. The effect of shape memory. - L .: Publishing house of Leningrad State University, 1987 .-- 216 p.

69. Parrinello M., Rahman A. // Phys. Rev. Lett. - 1980. - V. 45. - No. 14. - P. 1196–1199.

70. Parrinello M., Rahman A. // J. Appl. Phys. - 1981. - V. 52. - No. 12. - P. 7182–7187.

Conclusion

BCC alloys have unique properties as structural or functional materials; at present they are mainly produced by the ingenious technological processes. Of particular interest are the promising functional materials based on BCC alloys containing nanoscale elements in the structure. This is due to their structural features and properties in the pre-transitional low-stability state region. However, such materials manufactured artificially very often have poor resistance to external, primarily thermosilic, influences. Moreover, at present there are no methods for controlling their structure due to a lack of knowledge about the fundamental physical laws of their formation and behaviour.

It is important to emphasize that the pre-transitional low-stability state of a system is understood to mean its state near the structural-phase transformation in which anomalies of structure or properties are observed. It is this state that provides the functional properties of 'smart' materials, allowing one to vary the physicomechanical properties by changing external conditions. In the pre-transitional low-stability state, due to the low stability of the system to external influences, the concentration and role of structural defects significantly increase. Naturally, the traditionally understood structural defects under these specific conditions are already becoming integral elements of the structure, they interact with each other, and this interaction has a significant, if not decisive, effect on the structure and properties of the condensed system itself. It should be especially noted that the density of structural defects (defects in the traditional sense) is very high, therefore they cannot be considered as isolated defects, it is necessary to study the system of interacting defects under conditions of a low-stability state of the material. This in itself is far from a trivial task, especially taking into account the fact that not only the concentration of defects, but also their symmetry, the nature of the interaction, the plane of occurrence, the type and magnitude of the external influence and many other factors begin to play an important role. Given the low-stability the state of a condensed system to external conditions, the role of the interaction of structural defects acquires an especially important, and often the determining value for the structure.

When studying structural-phase transformations in the pre-transitional low-stability state of the system, the important objects of research are BCC metal alloys and intermetallic compounds, which have demonstrate peculiar behaviour in the vicinity of structural-phase transformations (for example, an alloy based on TiNi). This is due, first of all, to the fact that they have been studied for a long time and a large amount of experimental material has been accumulated with appropriate analysis and generalization. Of particular interest from the point of view of the choice of the object of study are those metals and alloys that are weakly resistant to external influences (temperature, load, composition change, alloying, etc.) for which there is a whole spectrum of structural states near the stability loss threshold; note that therse are either equilibrium states or those close to equilibrium. This means that these transformations themselves are either of the second kind or close to it, i.e., in such a low-stability pre-transitional state, a complex structural-phase state can be realized in the system, which continuously changes under the influence of varying the external and internal conditions.

The present book sets forth an attempt to look at a fairly wide number of different types of alloys with structural defects from a new physical point of view (based on a new understanding of the state of the system in which the traditional phase transition point has the form of an interval of values of the parameter controlling the transition. In such systems, the state of the material in this interval is structurally unstable with respect to the influence of small changes in the controlling parameter. Naturally, in such a thermodynamically and structurally weakly defined state, the interaction of structural defects on the thermodynamics of phase transformations and the structural state of the system.

At the beginning of the book, using the Metropolis Monte Carlo Algorithm, we considered the features of thermal cycling of pre-transitional low-stability structural-phase states of BCC alloys (using the example of the traditional CuZn alloy and NiAl intermetallic compound) in various situations: during one or several thermal cycles (during several consecutive heating-cooling cycles), in the presence of complexes of planar defects (shear and thermal antiphase boundaries), and the interaction of complexes of thermal antiphase boundaries (APB).

It is shown that in all cases, as a result of each heating and cooling cycle, a peculiar hysteresis is observed, the presence of which indicates the irreversibility of the processes implying a difference in the structural-phase states in the heating and cooling stages. It is concluded that the structural-phase transformations in the heating and cooling stages occur in different temperature ranges. In these intervals, the thermodynamic incentives for the realization of a structural-phase state are very small, which can be traced both on the dependences of the configurational energy, long-range and short-range order parameters, and on changes in the atomic structure and distributions of structural-phase states. Both ordered and disordered phases and a certain set of superstructural domains, are realized simultaneously. This means that in the vicinity of the disorder–order phase transition there are low-stability states.

The results of studies of several successive order–disorder and disorder–order phase transitions during several successive heating-cooling cycles demonstrate that for two successive heating-cooling cycles at the same temperature, the structural phase states differ both in the heating and at cooling stage. These differences are not revolutionary, but occur on each cycle, and decrease as the cycle number increases. Actually, the system undergoes training with the tendency for a certain steady-state sequence of structural-phase states.

It is shown that the influence of planar defects (antiphase boundaries) on the disordering process is significant up to the temperature of the structural-phase transformation. The most significant for the long-range order is the appearance of the defect itself, the difference in the type of antiphase boundaries and their plane of occurrence does not affect the behaviour of the long-range order with temperature. The type of antiphase boundaries significantly affects the structural and energy characteristics of the system at temperatures below the phase transformation temperature. Naturally, a system with structural defects is less ordered than a defect-free system. The presence of a defect contributes to the onset of disordering of the system at lower temperatures: a decrease in the order in the alloy begins in the case of thermal APBs at a lower temperature compared to the case of shear APBs. In an alloy with a complex of thermal APBs in the <100> direction, the first structural disorder in the CuZn alloy always appears near the Zn–Zn boundary.

In an alloy with a complex of shear APBs in the <110> direction, structural order disturbances at low temperatures are observed only in the regions of boundary intersection.

The presence of APBs affects the stability of the alloy during heating. A CuZn alloy without structural defects is more stable than an alloy with APBs. The disordering process is accompanied by blurring of the boundaries and their faceting.

The presence of a dual defect in the form of a pair of thermal APBs in an ordered BCC alloy with a *B*2 superstructure (using the CuZn alloy as an example) leads to significant structural-phase features of the system during the order–disorder transition compared to a defect-free system. The presence and nature of the observed features substantially depend both on the temperature and the distance between thermal APBs. In this case, Cu–Cu and Zn–Zn type boundaries differ both in the linear sizes and in the degree of ordering of the boundary regions. At the Cu–Cu interface, this region is smaller in linear dimensions and less ordered compared to those of the boundary region of the Zn–Zn interface. At low temperatures, the linear dimensions of the boundary disordered regions increase with increasing temperature in contrast to the background of a general decrease in the order in the system.

In the region of low-stability states of the system, the sizes of the boundary disordered regions are preserved: for the Cu–Cu type boundary it is about 10 interplanar distances, and for the Zn–Zn type boundary – about 12 of these. In this case, the ordering in these regions becomes close, blurring and faceting of APBs are observed, and the first disordered regions always appear near the Zn–Zn-type boundary.

Using the traditional CuZn alloy and NiAl intermetallic compound as an example, the influence of APB complexes (pairs of shear APBs and pairs of thermal APBs) on low-stability pre-transition states of BCC alloys is considered. It is shown that in the region of low-stability structural-phase states, the energy of formation of a complex of thermal APBs is higher than the energy of formation of a complex of shear APBs. The contribution of APBs to the disordering process is significant up to the temperature of the structural-phase transformation. The most significant for the long-range order is the appearance of the defect itself, the difference in the type of APBs and their plane of occurrence does not affect the behaviour of the long-range order with temperature. The type of APBs significantly affects the structural and energy characteristics of the system at temperatures below the phase transformation temperature. Naturally, a system with structural defects is less ordered than a defect-free system. The presence of a defect contributes to the onset of disordering of the system at lower temperatures: a decrease in the order in the alloy begins in the case of thermal APBs (TAPBs) at a lower temperature compared to the case of shear APBs. In the BCC system with the TAPB complex, the first ordering disturbances always appear near the antiphase boundary through which larger atoms (Zn–Zn in the CuZn or Al–Al alloy in the NiAl intermetallic) are adjacent.

In an alloy with a complex of shear APBs, ordering disturbances at low temperatures are observed only in the regions of boundary intersection. The presence of APBs affects the stability of the alloy during heating. It is shown that the disordering process is accompanied by blurring and faceting of the boundaries. From a comparative analysis of the features of disordering processes in the BCC system (the traditional CuZn alloy and NiAl intermetallic) with increasing temperature in the region of low-stability pre-transitional states it follows that while the order–disorder phase transition in the CuZn alloy occurs as a result of disordering in the system, then in the NiAl intermetallic compound long-range ordering occurs as a result of the structural-phase transformation. This means that the disordering process dominates in the traditional CuZn alloy, and in the NiAl intermetallic compound – structural-phase transformation.

In the study of NiAl intermetallic, special attention was paid to low-stability pre-transitional states. Since the intermetallic acid melts from an ordered state, hypothetical order–disorder transitions during heating and disorder–order transitions during cooling were considered. Using the Monte Carlo method on the example of the NiAl intermetallic compound of the Ni–Al system, it has been shown that irreversibility of processes is observed during thermal cycling in the course of structural-phase transformations in BCC intermetallic compounds. As a result of the heating and cooling cycle, a kind of hysteresis is observed, the presence of which indicates the irreversibility of the processes, which implies a difference in the structural-phase states in the heating and cooling stages. Analysis of the atomic and phase structure of the system in heating and cooling, i.e. in the process of hypothetical order–disorder and disorder–order phase transitions, confirmed the difference in structural phase states in the heating and cooling stages. It is shown that for the order–disorder transition to occur, the system needs to be slightly overheated relative to the traditionally understood phase transformation temperature, and for the disorder–order transition to occur, the system needs to be slightly supercooled relative to the same temperature.

During structural-phase transformations during step cooling in the alloy, the formation of two antiphase domains of superstructure $B2$ is observed. Analysis of the atomic and phase structure of the system during cooling showed the presence of elements corresponding to superparticle dislocations in the <100> plane and antiphase boundaries.

The analysis of the effect of the concentration of vacancies on the structural phase states and energy characteristics during heating and cooling showed that the presence and concentration of vacancies turn out to be significant factors in the region of pre-transitional low-stability structural phase states before the transformation. On the one hand, the presence of vacancies and their concentration do not affect the temperature ranges of structural-phase transformations, and on the other hand, they significantly affect the pre-transitional low-stability structural-phase states and the rate of diffusion processes. From the temperature behaviour of the short-range order parameter, it follows that the higher the concentration of vacancies (i.e., the higher the defectiveness of the system), the higher the tendency for atomic ordering to appear due to the intensification of diffusion processes. This, in turn, indicates an increase in the temperature of the onset of structural transformations with an increase in the concentration of defects in the alloy during cooling. An analysis of the temperature dependences of the long-range order parameter of the intermetallic compound allows us to conclude that an increase in the concentration of vacancies (i.e., an increase in the defectiveness of the alloy) leads to a natural result – a decrease in the long-range ordering in the system in the region of low-stability pre-transitional states and an increase in the temperature at which the transformation begins.

An analysis of the influence of the deviation of the atomic composition from the stoichiometric on the state of the intermetallic compound during cooling showed that the deviation turns out to be a significant factor in the region of pre-transitional low-stability structural-phase states before the transformation. The values of the short-range order parameter of non-stoichiometric alloys are much smaller in absolute value than the corresponding values of the stoichiometric intermetallic compound; therefore, the tendencies for the appearance of atomic ordering in non-stoichiometric intermetallic compounds are much smaller compared to the stoichiometric intermetallic compound. The behaviour of the temperature dependences of the long-range order parameter during cooling of alloys of non-stoichiometric compositions differs markedly from the behaviour of the corresponding dependence of the alloy of stoichiometric composition. When cooling the alloys of non-stoichiometric compositions, the establishment of a long-range order requires substantial supercooling, and the ordered phases appear at much lower temperatures. Moreover, the temperature dependence of the long-range order parameter of the Ni45Al55 alloy lies much lower than the corresponding curve of the Ni55Al45 alloy, which implies that the long-range order in these non-stoichiometric alloys is formed in different ways. It is noted that a deviation of the composition of the system from stoichiometric causes a significant refinement of ordered and disordered regions.

The study of the effect of grain size (model cell size) on the features of pre-transitional low-stability structural-phase states of the NiAl intermetallic in the region of structural-phase transformations during thermal cycling (heating and cooling) showed that a peculiar hysteresis is observed as a result of the heating and cooling cycle. With increasing grain size, the areas of hysteresis loops increase. With a small grain, the loop is closed, i.e. structural phase states after termination of the thermal cycle at low temperatures correspond to structural phase states at the same temperature until the thermal cycle is realized. With increasing grain size, the loop loses its closure, i.e. in the entire temperature range, the irreversibility of the processes and the difference in the structural phase states during heating and cooling at the same temperature begin to appear. Hence, the manifestation of the irreversibility of the processes and the difference in the structural-phase states during heating and cooling depend on the grain size of the alloy. With an increase in grain size, the temperature range of the transformation increases both during heating and cooling.

An analysis of changes in the short-range order parameter showed that in the region of low-stability pre-transitional states during heating, the alloy with the largest grain size shows the greatest tendency to ordering, and this tendency decreases during grain refinement. During cooling, the alloy with the smallest grain size shows the greatest tendency to ordering; when it is enlarged, this tendency decreases. As the grain size increases, the temperature range of the structural-phase transformations (the difference in the transformation temperatures during heating and cooling) increases.

An analysis of the temperature dependences of the long-range order parameters revealed that upon heating, the most complete long-range order is observed in the alloy with the largest grain size, and the least complete with the smallest grain size. To disorder the alloy with increasing grain size, an increasing overheating of the alloy is necessary. During cooling, the long-range ordering appears, first of all, in the alloy with fine grains. As the grain size increases, the temperature of the appearance of long-range order decreases, i.e. with increasing grain size, an ever increasing supercooling is required to realize the atomically ordered state of the system. The larger the grain size, the wider the temperature range of the structural-phase transformation.

The features of the formation of structural phase states during cooling, depending on the grain size (cell size of the model) indicate that the first ordered regions appear in the alloy with fine grains. As the grain size increases, the temperature of the appearance of long-range ordering decreases, i.e. with increasing grain size, an ever increasing supercooling is required to realize atomically ordered states of the system. Consideration of the influence of APB complexes (pairs of shear APBs in the <110> direction and pairs of thermal APBs in the <100> direction) on low-stability pre-transitional states of BCC intermetallic compounds (using the NiAl intermetallic as an example) showed that in the region of low-stability structural-phase states of the intermetallic intermetallic compound thermal APBs is higher than the energy of formation of a complex of shear APBs. The contribution of APBs to the disordering process is significant up to the temperature of the structural-phase transformation. The most significant for the long-range ordering in the system is the very appearance of a defect in the form of APB; the difference in the type of APBs and the plane of their occurrence does not affect the behaviour of the long-range ordering with temperature. Naturally, a system with structural defects is less ordered than a defect-free system. The presence of a defect in the form of APBs contributes to the onset of disordering of the system at lower temperatures: a decrease in the ordering in the alloy begins in the case of TAPBs at a lower temperature compared to the case of shear APBs. In an alloy with a complex of thermal APBs in the <100> direction, the first structural disordering in the NiAl alloy always appears near the Al–Al interface. In an alloy with a complex of shear APBs in the <110> direction, structural ordering disturbances at low temperatures are observed only in the regions of boundary intersection. The presence of antiphase boundaries affects the stability of the alloy during heating. It is shown that the disordering process is accompanied by blurring and faceting of the boundaries.

Using the example of the influence of point defects and their complexes on structural transformations in TiNi-based alloys, it is shown that in the low-stability state of a condensed system (in titanium nickelide these are the so-called pre-transitional states), the interaction of structural defects can have a significant effect on the features of structural phase transformations and nanoscale objects play a very important role in phase stability in the MT region.

In the pre-transitional low-stability state of a TiNi alloy of stoichiometric composition with the $B2$ superstructure (in the case under consideration, states with low elastic moduli), significant static displacements of atoms from crystal lattice sites take place before the transformation. As a result of this, even at a low concentration of defects, they interact with each other. In the final structure, atomic displacements in the vicinity of the defect are localized. The fields of atomic displacements in the vicinity of defects that do not lie in $\{110\}$-type planes prevent shuffling of the $\{1\bar{1}0\}$ planes in the <110> direction and the realization of the BCC→FCC martensitic transformation by this mechanism. Another variant of the development of martensitic transformation is not excluded. Thus, the interaction of defects located in the $\{111\}$ planes leads to the appearance of a long-period ω-like structure, which is formed by shuffling displacements of the <111> atomic rows. Thus, in the pre-martensitic state, when the system is at the boundary of its stability, the interaction of the defects of atomic displacement fields arising in the vicinity can influence the choice of a possible martensitic transition path.

Point defects caused by a deviation from stoichiometry and a disturbance of long-range order in the arrangement of atoms can affect the stability of the $B2$ structure and contribute to the transformation of the martensitic type into a ω-like structure. In this case, high-energy chains of point defects are formed in the <111> direction, which leads to an increase in the number of point defects in the martensitic phase. The latter can be significant in determining the return path of the martensitic transition.

Thus, using the example of studying the effect of point defects on structural transformations in titanium nickelide alloys, it is shown that in the low-stability state of a condensed system, the interaction of structural defects can have a significant effect on the features of structural phase transformations and nanoscale objects play a very important role in the stability of the crystal lattice in the field of MP.

When analyzing the effect of deformation on the temperature regions of martensitic transformations in the TiNi-based alloys, it was found that the concentration dependences of the onset temperature of the direct martensitic transformation M_s in stressed and unstressed specimens have different values and different functional dependences. This made it possible to reasonably assume that the observed effect is associated with the presence in the alloys of this class of low-stability pre-transitional states in the region of martensitic transformation, and the structure and properties of these states depend on previous thermomechanical influences and system states.

It was found that the temperature of the onset of the direct martensitic transformation M_s, found from the temperature dependence of the martensitic shear stresses, has a different value and a different functional dependence in comparison with the observed values and dependences in unstressed specimens and was obtained as a result of studying the temperature dependences of the electrical resistivity. To understand the results, a phenomenological approach based on the analysis of phase diagrams is used. From the analysis it follows that thermomechanical action should lead to an increase in the temperatures M_s and A_f. A logical assumption is made about the relationship between the observed effect and the presence of low-stability pre-transitional states in the martensitic transformation region in the alloys of this class, the structure and properties of these states depend on previous thermomechanical influences and system states.

The results of studies of physical properties in multicomponent Ti (Ni, Co, Mo) alloys with the effects of shape memory and the effect of annealing and thermal cycling on the intervals of martensitic transformations and on pre-martensitic low-stability states are presented. It has been established that thermal cycling through the MT range in microalloyed alloys leads to a slight decrease in the temperature of the onset of MT and a noticeable increase in the area under the temperature curve of electrical resistivity with saturation after the 20th cycle.

In alloys with a higher Co content, a slight increase in the peak area under the electrical resistance curve was found. It has been shown that in the $Ti_{48.94}Ni_{48.25}Co_{1.5}Mo_{0.31}$ alloy, the temperature range $R \rightarrow B19'$ is independent of thermal cycling. It was revealed that annealing at a temperature of 450°C leads to a change in the shape of the peaks in the temperature dependences of the electrical resistivity curves and causes a noticeable change in the characteristic MT temperatures.

It was found that the concentration dependences of the temperature of the onset of the direct martensitic transformation M_s in stressed and unstressed specimens have different values and different functional dependences. It is assumed that the observed effect is associated with the presence of low-stability pre-transitional states in the alloys of this class in the region of martensitic transformation, and the structure and properties of these states depend on previous thermomechanical influences and system states.

It is noted that the nonlinear nature of the temperature dependence of the electrical resistivity during thermal cycling and heat treatment in Ti (Ni, Co, Mo) alloys in the region preceding the martensitic transformation reflects the presence of a region of low-stability pre-transitional states near the stability loss boundary. The unusual effect of thermal cycling through the martensitic transformation region on the temperature dependences of the electrical resistivity curves in microalloyed multicomponent alloys based on titanium nickelide has been established, namely: thermal cycling leads to a noticeable increase in the area under the temperature curve of electrical resistivity, but no noticeable change in the temperature of the onset of the magnetic field was detected. In Ti (Ni, Co, Mo) alloys with a higher Co content, the character of the change in the curves of electrical resistivity versus temperature differs significantly from similar dependences in the microalloyed alloys. It is important that in the alloys studied, thermal cycling ends its significant effect on the MT by the 10th cycle.

The results obtained allow us to state that the effect of thermal cycling through the MT region in alloys with well-defined low-stability pre-martensitic states leads to the development of phase hardening with a wide range of defects. These defects also affect the mobility of interphase boundaries and can make a significant contribution to the increase in the number of the electron scattering centres.

When analyzing the relationship of structural defects and low-stability pre-transitional states, phase-structural transformations and stability of alloys, it was shown that in the low-stability state of a condensed system, the cooperative interaction of point defects can lead to their ordered arrangement, and the resulting static displacement fields can both stabilize the $B2$ structure and contribute to its instability and martensitic phase transition. In the presence of a certain type of defects in the $B2$ structure, the latter is unstable to displacement of the $\{111\}$ planes along the direction of the $<111>$ type. In the final structure, the displacement fields around the defects are localized, and the defects themselves organically fit into the structure of the formed phase. It should be noted that structural defects of the parent phase become natural elements of the structure of the final daughter phase. Structural defects in the low-stability state of the parent phase determine, in fact, the structure of the final daughter phase.

When studying the possibilities of the formation of a columnar structure of a relaxation type in BCC alloys, it was shown that the formation of a low-stability equilibrium columnar antiphase structure of a relaxation type in BCC systems is energetically quite possible. A necessary condition for its implementation in alloys with a basic $B2$ superstructure is the presence of anisotropy (for example, interatomic interaction). Moreover, a whole set of columnar structures of various sizes provides the energy advantage of the low-stability long-period state of relaxation type. It can be expected that under a small thermal force action the system will undergo structural-phase transformations, and at non-zero temperatures, a certain set of low-stability long-period states will coexist.

We studied low-stability pre-transitional states, order–disorder phase transitions and $B2$–$A1$ structural transformations in Cu–40 at.% Pd alloys, structural changes in the region of phase transitions directly in the temperature regions of the $B2$–$A1$ structural transition and order–disorder phase transitions. Based on the X-ray structural data, an analysis was made of the features of the $B2$ and $A1$ phases in the CuPd alloys near structural phase transitions and a picture of the behaviour of atomic vibrations in the crystal lattice was considered. It is shown that in the vicinity of structural-phase transformations in the CuPd alloys in the region of 40 at.% Pd low-stability states are formed in which a whole range of anomalous phenomena is observed (for example, anisotropy of atomic displacements, concentration inhomogeneities, stratification, heterophase fluctuations, nonlinearities in the dependences of the lattice parameters and long-range order parameters, etc.) that prepare the system for transformation.

The molecular dynamics method is used to study the features of the influence of planar structural defects (stacking faults) on structural rearrangements by the shear mechanism of BCC alloys in a low-stability state.

It is shown that in the low-stability pre-transitional state of the BCC system, the interaction of stacking faults can cause the transformation by the shear mechanism. At small distances between stacking faults, the transformation occurs by twinning. The possibility of the formation of polytype structures is demonstrated.

Using $B2$-alloys as examples, the relationship between the presence of planar structural defects (stacking faults) and the stability of alloys relative to structural-phase transformations in a low-stability state of systems has been studied. It was shown that during martensitic type transformations in BCC systems, the appearance of stacking faults introducing shifts of one part of the crystallite relative to the other in the {110} plane in the <110> direction leads to the thermodynamic profitability of the stacking fault structure, which slightly differs from the energy defect-free structure. The resulting structure corresponds to the global minimum of energy. In addition to this minimum, there are two local energy minima. The structure corresponding to the first minimum is formed by shuffling the {110} planes in the $<\overline{1}10>$ direction. The second local minimum corresponds to a metastable structure formed in addition to shuffling displacements of individual {110} planes in the $<\overline{1}10>$ direction by additional atomic displacements in the <110> direction.

Using the example of point and planar defects and their complexes, the inheritance of structural defects by the daughter phase during structural-phase transformations in the pre-transitional low-stability state of metal BCC systems has been studied.

During martensitic transformations for order defects, anti-structural atoms, and their simplest complexes, it was shown that in a number of cases, the number of point defects increases during martensitic transformations. In this case, the proportion of chemical energy accumulated by defects increases. The structure of a point defect inherited by martensite varies depending on the final reaction product. Certain types of defects in the $B2$ structure in martensite correspond to the appearance of high-energy linear disordered chains in the <111> direction. This can play a significant role in determining the path of reverse martensitic transformation.

For order defects, anti-structural atoms, and their simplest complexes, it has been shown that in a number of cases the number of point defects increases during martensitic transformation. In this case, the proportion of chemical energy accumulated by defects increases. Certain types of defects in the $B2$ structure in martensite correspond to the appearance of high-energy linear disordered chains in the <111> direction. This can play a significant role in determining the path of reverse martensitic transformation.

The structure of a point defect inherited by martensite varies depending on the final reaction product. The number of point order defects during martensitic transformation usually increases. High-energy defects are formed from linear chains oriented along the <111>-type direction. These defects favour an exactly a backward path of the reverse martensitic transformation.

A change in the number of defects during the martensitic transition leads to the accumulation by the defects of energy associated with the chemical contribution to the thermodynamic potential, since the number of pairs of atoms of the same type in the first coordination spheres changes. The displacements of atoms around defects in martensite are more localized than in the high-temperature phase. Therefore, the elastic energy associated with the non-chemical contribution from the defect decreases during the $B2 \rightarrow \omega$-like martensite transformation.

Index

F

factor
  Debye–Waller factor 190, 198, 199, 200
function
  Morse potential function 177

I

interface
  Cu–Cu interface 39, 47, 68, 75, 79
  Zn–Zn interface 36, 39, 47, 67, 75, 79

L

law
  Zen law 136, 137

M

martensite
  stress-induced martensite 147
method
  Metropolis Monte Carlo Method 5, 7
  Monte Carlo method 1, 12, 13, 24, 26, 40, 50, 91, 97, 111, 114, 118, 127, 128
  R.W. James method 197
model
  Debye model 190
Monte Carlo Metropolis algorithm 72, 77, 86

P

phase
  Hume-Rothery phase 2
phase transition
  disorder–order phase transition 1, 7, 8, 11, 12, 14, 16, 19, 25, 27, 31, 78
  order–disorder phase transition v, 2, 8, 11, 12, 16, 19, 20, 22, 24, 26, 29, 30, 31, 33, 34, 37, 40, 41, 42, 43, 44, 46, 47, 49, 50, 52, 53, 54, 60, 62, 66, 67, 69, 71, 72, 73, 74, 75, 80
potential 5, 7, 13, 27, 42, 52, 54, 72
  interatomic interaction potential 136
  Morse pair potential 5, 72
  Morse potential 7, 42, 54, 72
pre-transitional low-stability structural-phase states 1, 77

Printed in the United States
by Baker & Taylor Publisher Services

Printed in the United States
by Baker & Taylor Publisher Services